EXPLORING
Geospatial Technology

Rick Bunch
Elisabeth Nelson
James Nelson

Third Edition

Kendall Hunt
publishing company

Cover images provided by the author.

www.kendallhunt.com
Send all inquiries to:
4050 Westmark Drive
Dubuque, IA 52004-1840

Copyright © 2013, 2014, 2018 by Kendall Hunt Publishing Company

ISBN 978-1-5249-6244-9

Printed in the United States of America

Contents

Preface

We began teaching an introductory course in Geographic Information Science in 2004. During those early years, one of the most difficult issues we faced was finding a textbook that we considered appropriate for our course. We wanted one book that covered the basic theoretical concepts behind cartography, GIS, and remote sensing. We wanted those concepts presented in a language that was student-friendly to those outside the discipline of geography, and we wanted appropriate exercises to emphasize theoretical concepts in a hands-on learning environment. After several years of searching and coming up short, we decided to write our own. With our combined backgrounds in GIScience and computer technology, we were able to produce a book that presents the basic tenets that undergird the geospatial technologies of cartography, GIS, and remote sensing using an historical perspective as the link that ties the technologies both to each other and to GIScience. Our training also allowed us to include hands-on exercises at the end of each chapter. These exercises are designed to explore the concepts presented and take advantage of a variety of software packages, including ESRI's ArcGIS and Purdue University's *Multispec*.

Acknowledgments

We would like to thank the many students and colleagues at UNC-Greensboro who have offered their thoughts and suggestions on the topics and exercises covered in our text. We are especially indebted to those instructors and professors who have used our text and then taken the time to discuss with us their experiences. With the updates to the third edition, two students, Kaitlyn Jessee and Miguel Fernandez, provided valuable assistance by working through and editing each of the exercises. We extend our thanks to Kaitlyn and Miguel for their time and effort. Finally, thank you to those at Kendall-Hunt who have helped guide us through all the stages of writing and revising our text.

About the Authors

Rick Bunch is a Professor of Geography at the University of North Carolina Greensboro. His interests are in Geographic Information Systems (GIS) and the conceptualization, modeling and analysis of geographically referenced data. He serves on the editorial board for the journals of *Cartography and Geographic Information Science* and the *International Journal of Applied Geospatial Research*.

Elisabeth Nelson is a Professor Emerita of Geography at the University of North Carolina Greensboro. A former editor of *Cartography and Geographic Information Science*, her research specialty lies within the realm of cartography. She has been published in *Cartography and Geographic Information Science, Cartographica*, and *Cartographic Perspectives*. She enjoys teaching at all levels, and has a special interest in introducing students to the ideas and technologies of GIScience.

Jim Nelson is the Lab Director for the Department of Geography at the University of North Carolina Greensboro. With interests in cartography, urban planning, and GIScience, he excels at helping students navigate the complexities of geospatial technologies. He also enjoys challenging students of all levels with unique lab exercises.

Chapter 1

Geography and Geographic Information Science

Geographic information science (GIScience) is the fusion of multiple areas of study, but geography is its heart. As a core concept of our civilization, geography has often been characterized as the 'mother of all sciences'. Such a broad view has made it a difficult field to define succinctly, particularly as it has sought to keep pace with our rapidly changing world-view. First coined by *Eratosthenes*, head of the Great Library in Alexandria, Egypt, the term 'geography' is literally translated as 'earth writing' in Greek. As a definition, this might suggest that geography is more an amalgamation of all subjects related to the earth, from the physical to the cultural, rather than a discipline or subject in and of itself. What is it, then, that binds geographers and their various studies together?

One of the more influential arguments in favor of geography as a subject can be found in the classic work *The Four Traditions of Geography* (Pattison, 1990). Interestingly, this article doesn't attempt to define geography as much as it does to categorize it into related, but distinct traditions. In doing so, Pattison highlights that geographic studies are bound together into a single discipline via a common theme. The four traditions he outlines as central to geography are *Area Studies*, *Earth Science*, *Man-Land*, and *Spatial*. The Area Studies Tradition provides the home for world regional geography. Its focus is describing regions and their characteristics, highlighting both similarities and differences between areas and examining relationships between them. The Earth Science Tradition, on the other hand, is the home of physical geography. As such, it specializes in the physical features and natural processes of Earth, stressing Earth-Sun interactions and the core concepts of the Earth's four physical systems or 'spheres': atmosphere, biosphere, hydrosphere, and lithosphere. The tradition of Man-Land forms a bridge between Area Studies and Earth Science with its emphasis on the relationship between human societies and the environment. How humans influence and/or perceive the environment, as well as how nature may affect them (i.e., natural hazards) are themes examined under this tradition.

The common thread among these traditions, and the emphasis of the fourth and final Spatial Tradition, is the *exploration of spatial relationships*. The Spatial Tradition is devoted to developing tools and theoretical frameworks for studying spatial relationships in general; its focus is the use of spatial analysis to research and solve geographic problems. Thus, while it comprises a unique area of geographic study, the focus of this tradition and the use of the tools, frameworks, and analyses derived from it, is what all the traditions have in common. The key to all things geographical is that they all study and exploit spatial relationships.

Eratosthenes may have been the first to recognize geographical studies as unique, but the idea of a spatial perspective wasn't actually considered a formal category of knowledge until the 1700s. At that time, philosophers began to consider concepts such as distance, direction, and position as a specific perspective on the world. German philosopher *Immanuel Kant* was the first to identify geography as a unique form of knowledge. He considered it, along with history and the 'subject-matter' sciences (i.e., biology), to each be separate realms of intellectual inquiry (May, 1970). To grasp why Kant believed geography to be unique, visualize yourself studying the world through a *spatial lens*. When you do this, your focus shifts to one in which you are interested in *connections* – how the physical and cultural aspects of our planet relate to each other.

Kant, for example, would have argued that the biologist, the historian, and the geographer would all have different approaches to studying an historic site such as Guilford Courthouse National Military Park in Greensboro, North Carolina. The biologist, for instance, might focus on specific life forms associated with the park by planning and emphasizing bird conservation. The historian, on the other hand, would be more interested in the importance of the battle to the outcome of the Revolutionary War. These are two very different approaches to the study of the park, and although both might incorporate spatial components in their research, neither would most likely make the spatial component the primary emphasis of their work. The geographer is unique in that he or she would be most interested in the park's spatial relationship to surrounding geographic features. For example, Stine, et al. (2006) examined the locations of battle lines within the park in the context of Greensboro's current landscape, with the goal of enhancing a visitor's understanding of the spatial elements of the battle when they visit the park.

The Spatial Tradition and GIScience

Pattison's categorization of geographical studies into the four traditions clarifies how and why geography is a unique form of intellectual inquiry. It also highlights the role of GIScience in geography, which parallels the Spatial Tradition in its emphasis on developing tools and theoretical frameworks for studying spatial relationships. The tool of the Spatial Tradition and GIScience traditionally used in analog problem solving and data presentation has been the *map*. Defined by Robinson, et al. (1995:9) as a ". . . graphic representation of the geographical setting. . ." the map is an effective method of transmitting information and providing a venue for the analysis of spatial relationships. It is a well-suited vehicle for these uses because it deliberately simplifies real-world situations and presents results using abstract, but easily identified symbology.

Today's map is set in a dynamic, computerized setting that has provided cartographers and geographic information scientists (GIScientists) with the opportunities to design new types of maps and to allow users to explore and interact with the spatial data comprising them. The addition of interaction encourages spatial and visual thinking processes as users explore spatial data relationships and patterns (Menno-Kraak, 2006). If anything, then, the map has become even more powerful and indispensable to geography as the discipline has moved into the digital environment.

Characterizing the map as capturing spatial relationships in graphic form highlights its use of *spatial structure*, which emphasizes location and attributes, two key characteristics of any map (Robinson, et al. 1995). *Location* is a place characteristic defined by using coordinate systems to designate a place's position on Earth. *Attributes* represent a name, label, measurement, or value associated with a location or pair of coordinates (Figure 1.1). The combination of location and attributes comprises *geospatial data*. It is from these two basic data elements that you can begin to explore relationships: relationships between locations, such as distance; relationships between attributes, like income and education in Memphis, TN; or relationships between an attribute at multiple locations, such as varying rainfall amounts between Memphis and New Orleans. The result is a study of patterns and interactions – *spatial analysis* – that occurs over a given geographic region.

With the realm of the map now being the digital environment, the processes of collecting, analyzing, and visualizing geospatial data are all shaped by constant changes in computer technology. The initial movement of the mapping process into this environment has its roots in the early 1960s. It was during this timeframe that

FIGURE 1.1
Location and attributes.
Data: *Made with Natural Earth. Free vector and raster map data @ naturalearthdata.com.*
Source: Elisabeth Nelson

Location: 35°N, 90°W
Attribute: Memphis
Attribute: Tennessee

Location: 30°N, 90°W
Attribute: New Orleans
Attribute: Louisiana

Geographic Information Systems (GIS), envisioned as a "digital toolbox for handling maps" (Goodchild, 2004:709), were first implemented with the goal of alleviating the more monotonous and often expensive tasks cartographers and surveyors performed when working with spatial data. Considered one of the core systems under the current geospatial technology umbrella, GIS is a computerized information system designed specifically to capture, manage, analyze, and present spatial data.

The maps produced in a GIS are often described as being composed of layers of information created from different types of spatial information. These layers might include water features, parcels, elevation, roads, etc. (Figure 1.2). When the layers take the form of a static printed map, you can visually examine the spatial relationships between them. You might examine a county map, for example, to locate and count the number of elementary schools. You could also visually estimate the clustering, or distances between these schools and perhaps infer from that the population densities of the areas around them. Imagine now that these layers of information can be accessed and displayed on your home computer. Within the digital environment of a GIS, you can use built-in functions to have the computer analyze these layers. The GIS can tell you not only how many elementary schools there are within the county, but also how many are within, say, seven miles of the largest city in the county, which ones are within specific census tract areas, and the demographics of the populations within those tracts.

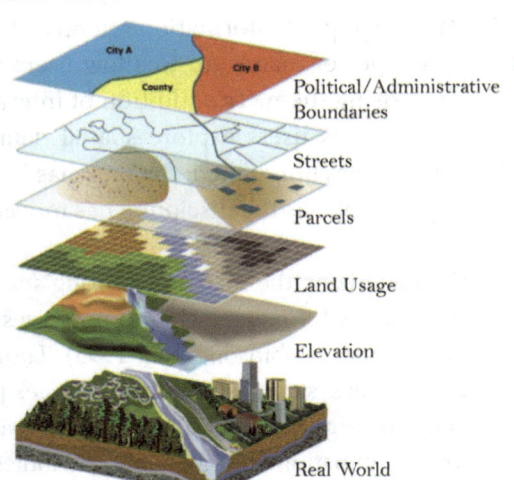

FIGURE 1.2
GIS conceptualization of spatial features as layers of information. Graphic courtesy of the *National Weather Service (https://www.weather.gov/bmx/gis).*

These types of built-in functions took time to develop; the GIS of today has benefitted from a 50-year evolutionary process. From the beginning, those involved in the development of these systems faced both computational as well as conceptual challenges. As the field matured, some began to question whether there might be a science behind the technology comprising an academic subdiscipline. In 1992, Michael Goodchild of the University of California at Santa Barbara published one of the first lists of GIS-related research priorities. The title of the paper was *Geographical Information Science,* and while a definition of GIScience was not offered, Goodchild did present the academic community with the components he considered the scope of that science. These included *data collection, data capture, spatial statistics, data modeling, data structures, data display, analytical tools,* and *managerial and ethical issues.* One, if not the first, definition of GIScience, came from the University Consortium for Geographic Information Science (UCGIS). This entity was formed in 1994 by a group of academics who came together to form the first committee for basic GIS research (Mark, 2003). They defined their group, and thus, indirectly, GIScience, as "*. . . dedicated to the development and use of theories, methods, technology, and data for understanding geographic processes, relationships, and patterns*" (UCGIS, 2002).

GIScience Today

In 2006, the UCGIS sought to formalize the difference between geographic information systems (GIS) and geographic information science (GIScience). Early GIS, along with cartography and remote sensing, were technologies that initially evolved separately, both in product design and research agendas. As they have all moved to the digital environment, however, they have begun to ". . . serve overlapping applications, and face similar issues of representation, database design, accuracy, and visualization" (Goodchild, 2004:710). GIScience is the subdiscipline that addresses these issues and the principles of spatial information processing under which current GIS, cartographic and remote sensing technologies operate.

The UCGIS (2006) now considers geospatial technology and GIScience to be two of three interrelated sub-domains of *Geographic Information Science and Technology (GIS&T)*, a broader framework proposed to address the continually evolving discipline of GIScience. The third sub-domain is *Applications of GIS&T*. Within this framework, which is also often referred to as *geospatial technologies* or *geotechnology*, GIScience represents the conceptual underpinnings of GIS, while geospatial technology refers to the set of specialized technologies that handle spatial data. Applications cover the myriad of uses of these technologies in different sectors of employment and across different disciplines. It is noteworthy that the U.S. Department of Labor named *geotechnology* as ". . . one of the three most important emerging and evolving fields, along with nanotechnology and biotechnology" in 2004 (Gewin 2004).

GIS&T, then, is presented as an integrative realm, one that ties research together across many related disciplines and expands how we think about GIScience. Two events resulted from this change in perspective. First, UCGIS called for a greater need to teach students the concepts that form the foundation of GIS&T, as well as the expertise in how to apply these technologies, as acknowledged by the U.S. Department of Labor. Secondly, this group has formally expanded the view of what it means to "do" GIS. GIS is now one component of a much larger knowledge realm, and understanding each component in this realm is essential to fulfilling the need for this pool of labor the government has forecast.

FIGURE 1.3
Marble's Competency Levels for Degree Programs, 1998. Source: Elisabeth Nelson

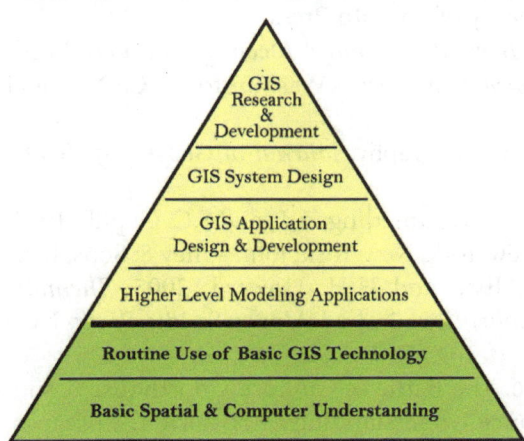

Exploring Geotechnology emphasizes the conceptual understanding required to work with GIS, cartography, and remote sensing properly at a basic level. Marble has (1999:31) argued that not only is GIScience education lacking in depth, but also, particularly at lower levels, in breadth. He notes *"Existing GIS education fails to provide the background in GIScience that is necessary to meet the needs of either users of GIScience technology or of the scientific community engaged in basic*

GIScience research and development." Marble (1998) even went so far as to identify competency levels that he felt degree programs should provide (Figure 1.3). The base of the pyramid, *Basic Spatial and Computer Understanding*, is the focus here. The pervasiveness of GIS across disciplines, he argues, demands that geography step up and critically engage students at this level before allowing them to move upward. Primary concerns here include learning how to *think spatially* and how to strike a balance between learning complex software and forming an understanding of the foundational concepts common to the geospatial technologies. Both are crucial to minimizing errors and efficiently using GIS at more advanced levels.

Additional Readings

The American Heritage Dictionary of the English Language. 1993. 3rd ed. Boston, Massachusetts: Houghton Mifflin.

Bonnett, A. 2008. *What is Geography?* Los Angeles, California: Sage Publications.

Brown, L.A. 1977. *The Story of Maps.* New York, New York: Dover Publications, Inc.

Campbell, J. 1998. *Map Use & Analysis.* 3rd ed. Boston, Massachusetts: WCB McGraw-Hill.

Clarke, K.C. 2003. *Getting Started with Geographic Information Systems.* 4th ed. Upper Saddle River, New Jersey: Pearson Education, Inc.

Dent, B.D., Torguson, J.S., and T.W. Hodler. 2009. *Cartography: Thematic Map Design.* 6th ed. New York, New York: McGraw-Hill.

DiBiase, D., DeMers, M., Johnson, A., Kemp, K., Luck, A.T., Plewe, B., and E. Wentz. (Eds.). 2006. *Geographic Information Science and Technology Body of Knowledge.* 1st ed. Washington, D.C: Association of American Geographers.

Gewin, V. 2004. Mapping Opportunities. *Nature,* 427:376-377.

Goodchild, M. 1992. Geographical Information Science. *International Journal of Geographical Information Systems,* 6(1): 31-45.

Marble, D.F. 1998. Rebuilding the Top of the Pyramid. *ArcNews,* 20(1): 28-29.

Marble, D.F. 1999. Developing a Model, Multipath Curriculum for GIScience. *ArcNews,* 21(1): 1, 31.

Mark, D. 2003. Geographic Information Science: Defining the Field. In M. Duckham, M. Goodchild, M. Worboys (eds.) *Foundations of Geographic Information Science. New York, New York: Taylor and Francis.*

May, J.A. 1970. *Kant's Concepts of Geography and Its Relation to Recent Geographical Thought.* Toronto, Canada: University of Toronto Press.

National Academy of Sciences. 2006. *Beyond Mapping: Meeting National Needs Through Enhanced Geographic Information Science.* Washington D.C.: National Academies Press.

Pattison, W. 1990. The Four Traditions of Geography, *Journal of Geography.* Sept.-Oct.: 202-206.

Robinson, A.H., Morrison, J.L., Muehrcke, P.C., Kimerling, A.J., and S.C. Guptill. 1995. *Elements of Cartography.* 6th ed. New York, New York: John Wiley & Sons, Inc.

Slocum, T.A., McMaster, R.B., Kessler, F.C., and H.H. Howard. 2005. *Thematic Cartography and Geographic Visualization.* 2nd ed. Upper Saddle River, New Jersey: Pearson Education, Inc.

Stine, R., Nelson, E. and M. Swaim. 2006. The Battle of Guilford Courthouse: Using Cartographic Animation to Enhance Understanding of Historic Landscapes. *Research in Geographic Education,* v. 8: 73-89.

Exercise 1

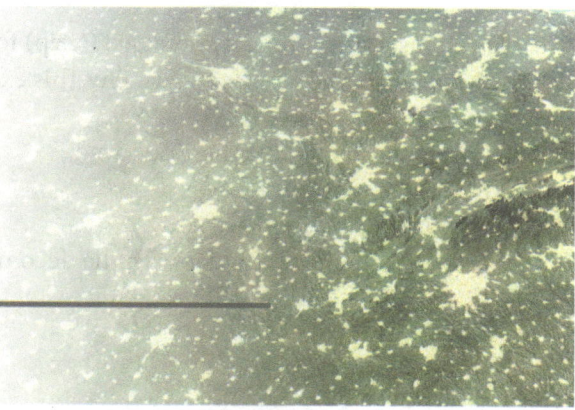

Introducing ArcMap

In Exercise 1, you will explore the environment of **ArcMap**. You will create an **ArcMap** map document (*.mxd), and then add and symbolize map layers. You will also explore the map you create and its associated attribute data using some of **ArcMap's** basic tools.

OBJECTIVES

- Identify the major components of the ArcGIS Interface
- Learn the functions of the basic tools located on the *Tools* toolbar
- Use the *Standard* toolbar tools to add data layers and control map scale
- Identify the basic elements in the *Table of Contents (TOC)*
- Use the *TOC* to control layer visibility, layer order, and symbol properties
- Establish folder connections using *Catalog*

SOFTWARE INFORMATION

Introducing ArcMap uses ESRI's **ArcGIS 10.6** software. **ArcGIS** is a commercial GIS package, available in geospatial labs affiliated with schools that have a campus-wide site license for the software. Instructors at these campuses may also request 1-year student versions of the software at http://www.esri.com/landing-pages/education-promo.

DATA

The data for the exercises is available from the Kendall Hunt Student Ancillary site. See the inside front cover for access information. You may also be directed to download the data from a different location by your instructor.

1. Download the data as instructed by your instructor
2. Save the file to a location where you have read/write privileges (USB key, home computer, class server)

The file you just saved is in a compressed (*.zip) format, and was created using **7-zip** freeware (http://www.7- zip.org). To use the data for this exercise, you must decompress it using the same software.

3. Locate the zipped file that you saved
4. Right-click the file
5. Choose *7-Zip - Extract Files*
6. Click *OK* to create a folder with the decompressed data

The ArcMap Interface

CONCEPT

To use **ArcMap** efficiently, you must be able to navigate its graphical user interface (GUI). This interface can be dissected into four general areas: *Toolbars, Table of Contents (TOC), Data View,* and the *Dock*. Their general layout within the interface is shown in Figure 1.1.

FIGURE 1.1
Used with permission. Copyright © 2018 Esri (Environmental Systems Research Institute), ArcGIS, ArcMap, ArcCatalog, United States Department of the Census; naturalearthdata.com All rights reserved.

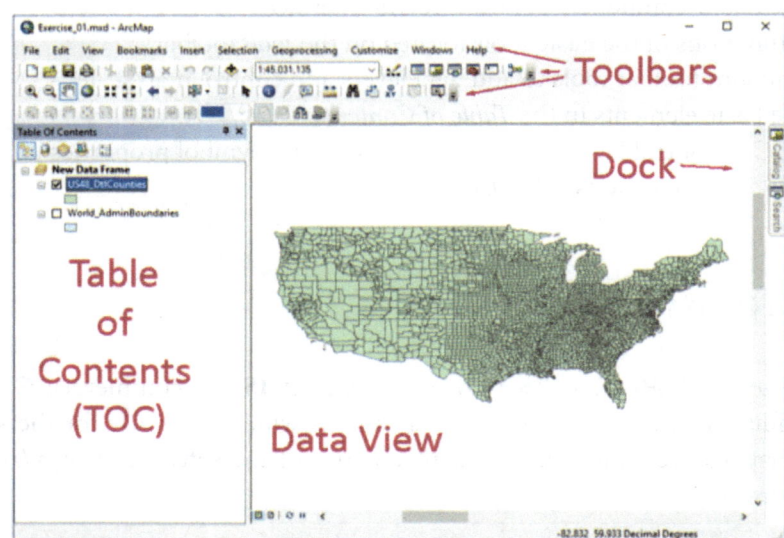

1. Start **ArcMap**. You can access ArcMap using several different methods, depending on your environment. You can start by clicking the Windows start menu icon at the bottom left corner of your screen, then navigate to the ArcMap icon in the ArcGIS folder, or search for ArcMap in the search window by typing "arcmap".
2. In the "*ArcMap - Getting Started*" dialog box, click *Browse for more* under *Existing Maps* (Figure 1.2)

FIGURE 1.2

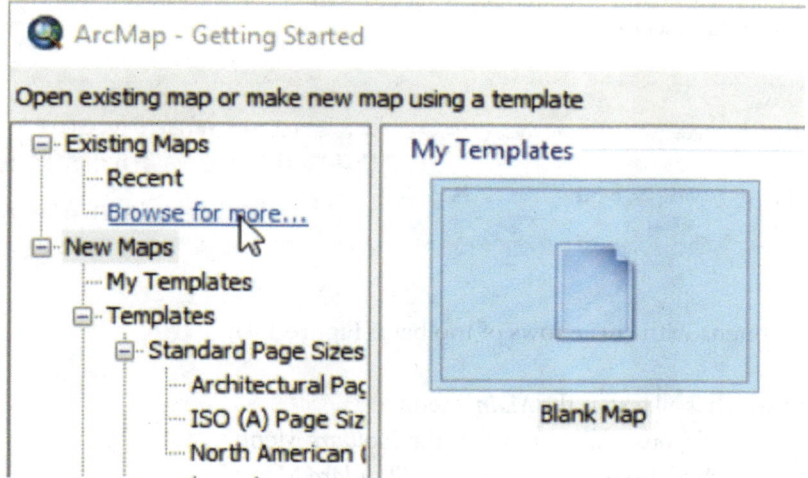

FIGURE 1.3

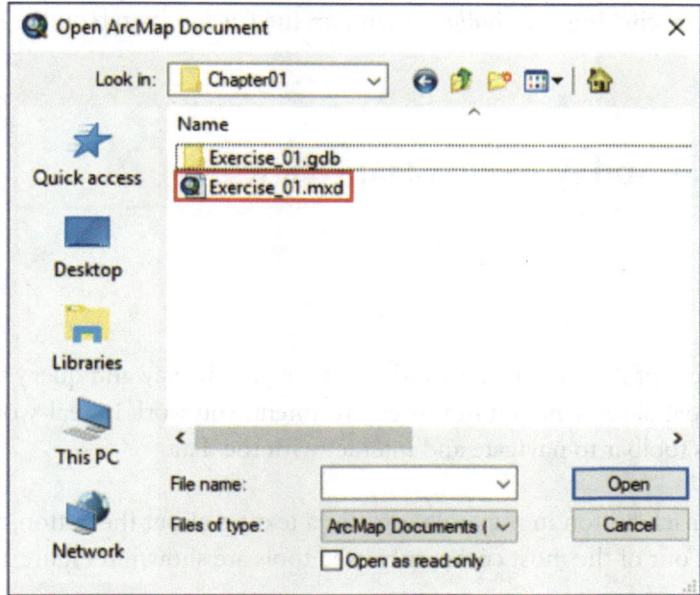

3. Navigate to the location where you saved the *Chapter01* data folder
4. In the folder, double-click **Exercise_01.mxd** to open the map document (Figure 1.3). **ArcMap** should open, with a map of the counties of the United States displayed, similar to Figure 1.1.

FIGURE 1.4

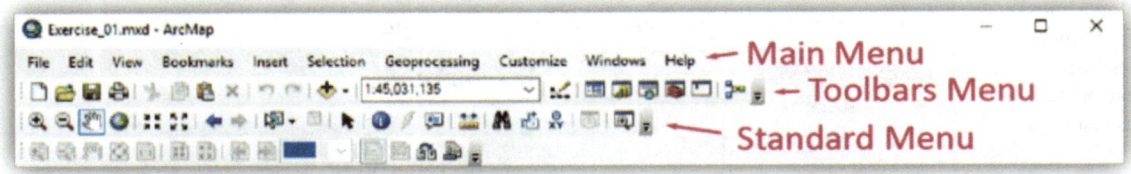

By default, **ArcMap** opens with **three** rows of toolbars (Figure 1.4):

- The first row, which is all text, is the *Main* Menu
- The second row is all icons, and is known as the *Toolbars* Menu
- The third row, which is also mostly icons, is the *Standard* Menu

Helpful Hint: You can hover your cursor over a button to get a hint about its function.

Other toolbars may or may not be displayed, depending on your computer setup. The wide variety of toolbar options can be accessed by clicking the *Toolbars* menu on the *Customize* tab on the *Main Menu.*

The Tools Toolbar and ArcMap's Data View

CONCEPT

The Data View component of **ArcMap** is the window where you display and query your spatial data. When you interact with your spatial information in this environment, you work in real-world coordinates and you use the tools in the *Tools* toolbar to navigate and interact with the data.

Hover your cursor over each button in the toolbar to see a text tip about the button, and click each button to enable its functionality. Four of the most commonly used tools are shown in Figure 1.5.

1. Select the *Zoom In* tool – click and drag your cursor in the *Data View* to zoom in to an area of interest
2. Select the *Zoom Out* tool– click in the *Data View* to zoom out
3. Select the *Pan* tool – click and drag to move to an area of interest in the *Data View*
4. Select the *Zoom to Full Extent* tool– click in the *Data View* to zoom out to the full extent of the map

EXPLORING GEOSPATIAL TECHNOLOGY

FIGURE 1.5

Zoom In

Zoom Out

Pan

Zoom to
Full Extent

5. Click on the *Bookmarks* tab on the *Main Menu*, then click "US_48"
6. Zoom in to the state of *Maine (ME)* in the top right-hand portion of the map

FIGURE 1.6

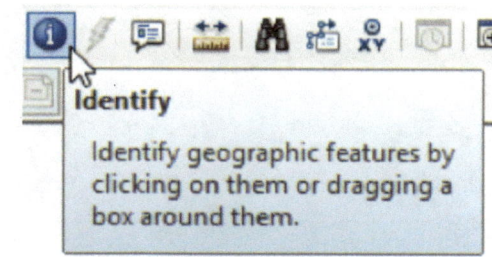

Identify

Identify geographic features by clicking on them or dragging a box around them.

7. Click on the *Identify* tool, and then click the northernmost or top-most county in *Maine* (Figure 1.7)

A new window will open, showing you the data that has been collected and stored for this county, including the county name.

FIGURE 1.7

Overview map of the contiguous U.S. Made with Natural Earth. Free vector and raster map data @ naturalearthdata.com. Source: Elisabeth Nelson

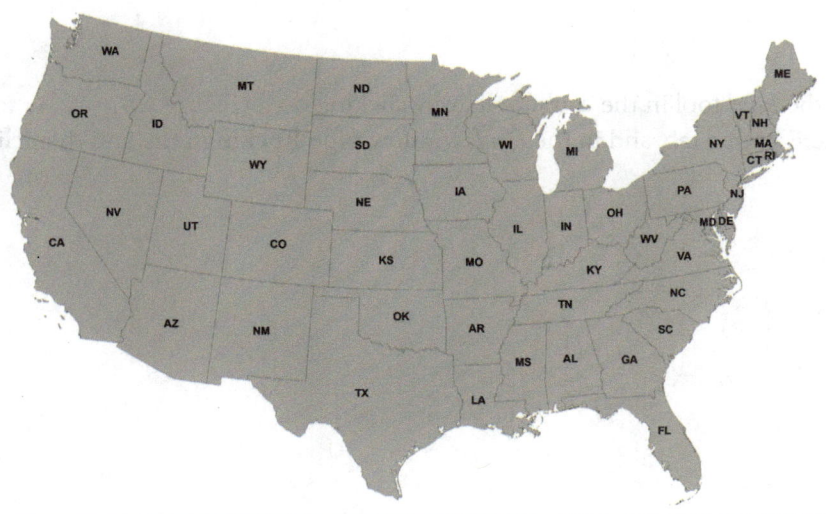

QUESTION 1.1

What is the name of this county?

    a. Washington
    b. Somerset
    c. Penobscot
    d. Aroostook

8. Close the *Identify* window by clicking the X in the top, right-hand corner of the *Identify* window
9. Select the *Pan* tool and navigate to the state of *Texas (TX)*
10. Use the *Identify* tool to identify the southernmost county in *Texas*

QUESTION 1.2

What is the name of this county?

    a. Zapata
    b. Cameron
    c. Hidalgo
    d. Willacy

11. Close the *Identify* window

FIGURE 1.8

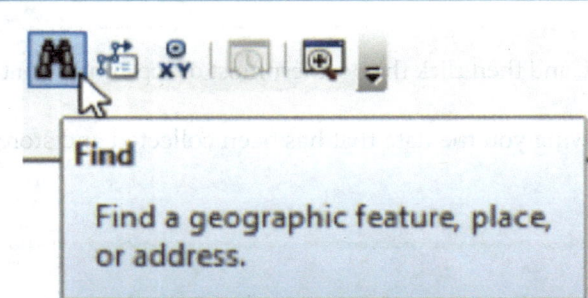

12. Activate the *Find* tool in the *Tools* toolbar by clicking on it (Figure 1.8)
13. Select the *Features:* tab, and in the *Find:* window, type **Rockingham** and then click *Find* (Figure 1.9)

                                                   EXPLORING GEOSPATIAL TECHNOLOGY

FIGURE 1.9

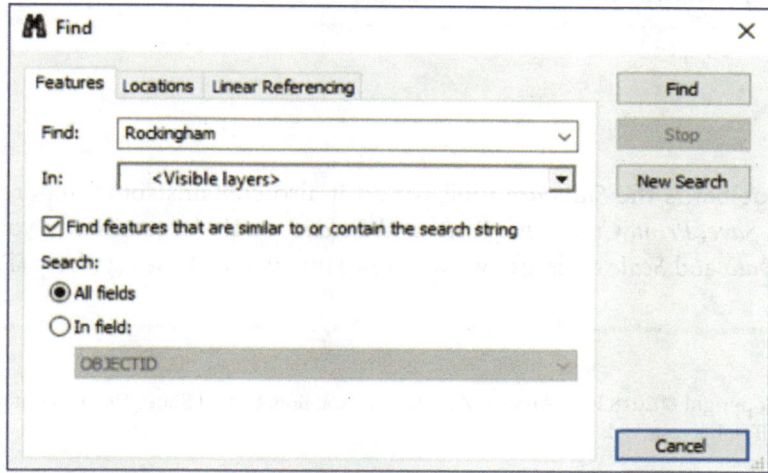

The bottom of the *Find* window shows all of the counties in this map layer with the value of *Rockingham*. Drag the dialog box out of the way of the map so you can see both it and the *Find* box.

14. Right-click the top value and choose *Identify*
15. Scroll through the information provided to answer the following question

QUESTION 1.3

This county is part of what state?

     a. Virginia

     b. Utah

     c. South Carolina

     d. Indiana

16. Use the *Find* and *Identify* tools in the steps above to answer the following question:

QUESTION 1.4

The second and third counties listed using the *Find* tool are part of which states?

     a. Oregon and Oklahoma

     b. North Carolina and North Dakota

     c. Georgia and Vermont

     d. North Carolina and New Hampshire

17. Close the *Find* and *Identify* windows

The Standard Toolbar and the Catalog Window

CONCEPT

The second, lower toolbar is the *Standard* toolbar, and it also contains some important features. Some of these, such as *Open, Save, Print, Cut, Copy, Paste,* and *Delete*, should be familiar to you from other software packages. The *Add Data* and *Scale* tools are two new ones that you will use regularly (Figure 1.10).

FIGURE 1.10

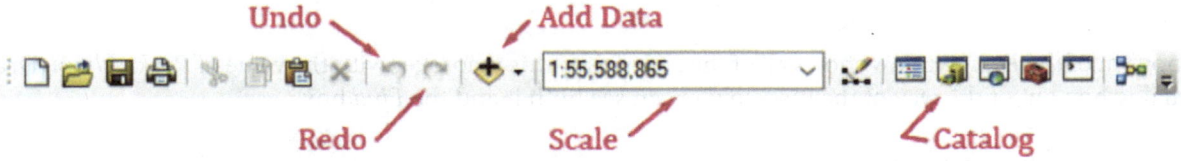

1. Scroll along the *Standard* toolbar to get text tips

The Catalog window is used to organize and manage your data layers, map documents, and analytical tools, among other elements. It organizes your documents and spatial data using file folders and geodatabases, and presents the results using a tree view similar to what you see when working in *Windows Explorer.*

2. If the *Catalog* window is not open, click on the *Catalog* icon on the Standard toolbar, or the Catalog tab on the right side of the **ArcMap** interface to open it
3. Click the pushpin in the upper right corner of the *Catalog* window to freeze the window in place (Figure 1.11)

FIGURE 1.11

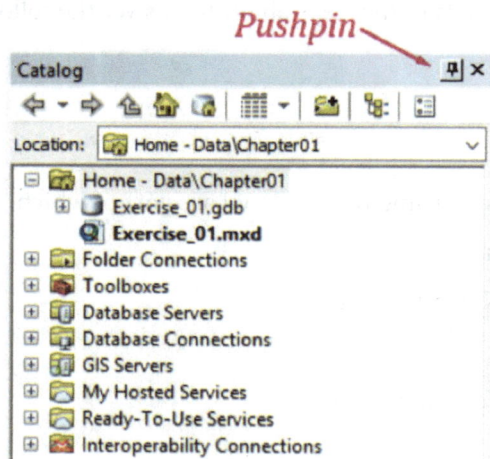

One of the primary tasks for which the *Catalog* window is used is to establish connections to folders of data you wish to use in your map documents.

4. Right-click *Folder Connections* and choose *Connect to Folder*
5. Navigate to the location where you saved the *Chapter01* folder of data for this exercise
6. Click once to select the *Chapter01* folder, then click *OK* to create a folder connection that will allow you to access the data inside the folder
7. Click the + sign to the left of the folder to see its contents. You should see a geodatabase named *Exercise_01*
8. Click the + sign to the left of the geodatabase to see the actual spatial data files you will use in this exercise
9. Click and drag the *World_Cities* feature class from the *Catalog* window to your *Data View* window

In the scale drop-down box on the Toolbars menu, the current scale is displayed as a representative fraction (RF).

10. Click the RF to select it
11. Type **1:20,000,000** and press the *Enter* key to select this as the new map scale
12. Now, type **1:10,000,000** and press the *Enter* key

QUESTION 1.5

When you change the map scale from 1:20,000,000 to 1:10,000,000, the:

a. map zooms in and the map scale gets larger
b. map zooms in and the map scale gets smaller
c. map zooms out and map scale gets larger
d. map zooms out and the map scale gets smaller

The Table of Contents (TOC)

CONCEPT

The *TOC* is where you will manipulate the map document's maps and their spatial data layers. An individual map is represented by a *data frame*; in this map document the default data frame is called *New Data Frame*.

Under the data frame, the layers that comprise the map are listed, and the order in which they are listed affects the look of the map. The layers listed at the top of the *TOC* draw on top of those that reside below them.

The *TOC* is also where you will find what each feature represents and where you set the symbol properties used to represent the feature on the map. In addition, you can make each layer visible or invisible and view the path that will take you to where the map layer is physically stored on your machine.

1. Hover your cursor over the icons directly under the *Table of Contents* heading
2. Click on the *List by Source* button (Figure 1.12)

The data layers listed in the *TOC* are individual files, but *the files are not stored as part of the map document.* Instead, the map document stores a link, or path, to the data file. *List by Source* lists the layers by the computer path to their data source.

QUESTION 1.6

What is the path to the location where your data is stored?

_____

3. Click on the *List by Drawing Order* button (Figure 1.12)

List by Drawing Order lists the features in the order in which they will draw. You can reorder these layers by clicking and dragging them up and down the list.

4. Right-click the *Layers* data frame, and then click *Turn all Layers On*
5. Click and drag the bottom layer (*World_AdminBoundaries*) to the top of the *TOC*

QUESTION 1.7

When you click on a data layer in the *TOC* and drag it to the top of the list of data layers:

a. It never draws
b. It draws in a random order in relation to the other data layers
c. It draws first, and is covered by all the other layers
d. It draws last, covering everything below it

More on the Data View

CONCEPT

The *Data View* is where you examine, edit, and analyze the spatial data comprising your map or *active* data frame (the active data frame is the one with its name in **bold**). It only displays the data layers – other map elements, such as the title, legend, scale, etc., are hidden from view in the data view.

Through the *Data View*, you can visualize the spatial data in many ways. **ArcMap** allows you to modify the symbolization technique, as well as how the data is classified. The initial default view **ArcMap** provides uses a single symbol for each data layer.

1. Right-click in the *Data View* and select *Data Frame Properties* from the context menu

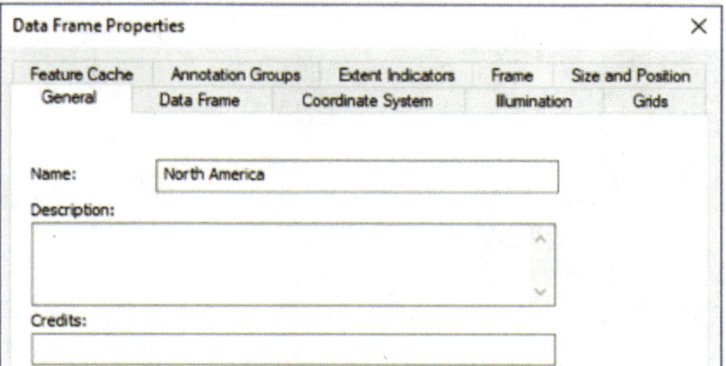

Data Frame Properties provides a set of tabs that allow you to change certain properties that affect every data layer in the data frame (Figure 1.13).

2. Select the *General* tab and in the *Name* box change the name of your data frame or map to **North America**
3. Click *Apply* (if a warning box pops up, close the warning)
4. Select the *Frame* tab and click the drop-down arrow under *Background*
5. Select *Blue* and click *Apply* to add a background color to your map

Probably the most important tab for the data frame is the *Coordinate System*. Every data frame represents some part of the world in two dimensions, so each one will have a specific coordinate system for displaying that area.

By default, *the coordinate system is the same as the first data layer you add to the data frame*. Many times, however, you will want to modify that system to meet certain requirements for your map.

6. Select the *Coordinate System* tab (Figure 1.14)
7. In the top scroll box, scroll to find the *Projected Coordinate Systems* folder
8. Click the + sign to expand the *Projected Coordinate Systems* folder
9. Click the + sign to expand the *Continental* folder, then do the same for the *North America* folder
10. Click to select *WGS 1984 Canada Atlas LCC*
11. Click *Apply*
12. If a Warning dialog pops us, click *Yes* to move through it – this will be addressed later
13. Click *OK*

FIGURE 1.14

To change specific symbol colors and styles, right-click on each individual data layer and select *Properties*.

14. First, click and drag the *World_AdminBoundaries* back to the bottom of the *TOC* list
15. Right-click the *World_Cities* layer and select *Properties*
16. Select the *Symbology* tab
17. To change the color, size, and/or shape of the symbol, click on it
18. Click on the color drop-down box and hover over the colors to find and select *Sage Dust* (green family, 4th row from the bottom)
19. Click *OK*, then *OK* again
20. In the *TOC*, uncheck the box next to *US48_DtlCounties* to make the layer invisible
21. Click the colored box under *World_AdminBoundaries* to bring up the *Symbol Selector* dialog
22. Scroll down and select the *Grey* box
23. Click the drop-down arrow next to the *Outline Color* box and choose *Gray 50%*
24. Click *OK*

EXPLORING GEOSPATIAL TECHNOLOGY

SAVE YOUR MAP DOCUMENT

1. From the *Bookmark* menu, choose *US_48* to automatically go to the U.S. lower 48
2. From the *File* menu, choose *File Save As*
3. Name your file using this convention: *lastname_Exercise01*
4. Save your map to a location, such as a class server, that both you and your instructor can access
5. Now, from the *File* menu, choose *Export Map*
6. In the *Save In* drop-down box at the top of the dialog, navigate to the same location that you saved the original map document. Under the *Save as Type* drop-down box, choose *PDF*
7. Name your file to match the name of the original map document minus its extension, then click *Save*
8. Double-click the *Computer* icon your desktop
9. Navigate to the location where you think you saved your file and verify that it is there

Note: To receive full credit during the grading process, the map you export should reflect the latest changes as directed by the exercise.

QUESTION 1.8

To change your map's coordinate system, you use the:

a. Layer's Properties
b. Symbology tab
c. Data frame's Properties
d. General tab

QUESTION 1.9

The primary purpose of the *Data View* component of **ArcMap's** interface is to allow you to work with map layout elements, such as titles, north arrows, and scale bars, and to arrange these on a page in relation to a data frame.

a. True
b. False

32. Click the pushpin icon in the top right on the *Catalog* window to hide the window
33. From the *File* menu, choose *Save*
34. Exit **ArcMap**
35. If you are on a machine in a lab and are finished with your computer session, log off and/or restart your machine before you leave.

Chapter 2

Spatial Representation

In Chapter 1, we defined a map as a "... graphic representation of the geographical setting..." (Robinson, et al., 1995:9). In this setting, the word *representation* may refer to a likeness of the environment, such as the image of the United States in Figure 2.1, or to an abstract concept, such as the spatial variation of those 35 and older who are enrolled in school (Figure 2.2). In both instances, the result is a method of communication that relies on one of four fundamental forms of intelligence. The method is *graphicacy*, and the type of intelligence upon which it relies is known as *visuospatial intelligence*. Communicating via graphicacy requires that we be able to generate and understand graphics and images, including photographs, diagrams, maps, charts and plans, among others.

FIGURE 2.1
Relief map of the United States.

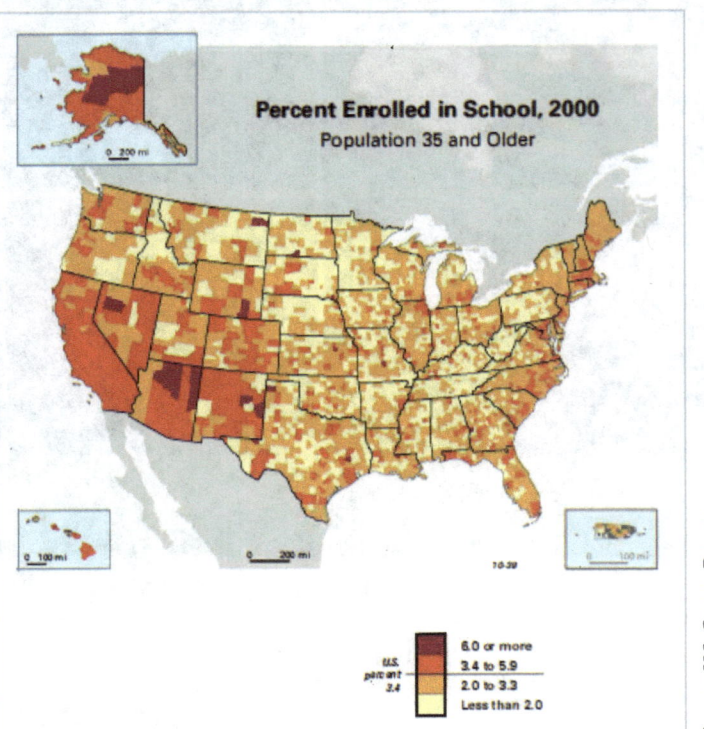

Percent Enrolled in School, 2000
Population 35 and Older

6.0 or more
3.4 to 5.9
2.0 to 3.3
Less than 2.0

U.S. percent 3.4

Source: U.S. Census Bureau

FIGURE 2.2
Percent of population 35 and older enrolled in school in 2000. Suchan, T., Perry, M. Fitzsimmons, J., Juhn, A. Tait, A., and C. Brewer. *Census Atlas of the United States*, Series CENSR-29, U.S. Census Bureau, Washington, DC, 2007.

As a vehicle for communication, graphicacy is surely one of the earliest methods humans used; unfortunately, its power often goes undetected in American education, where developments in *verbal*, *social*, and *numerical* intelligences (and the corresponding communication methods of *literacy*, *articulacy* and *numeracy*) dominate (Aldrich and Sheppard, 2000). This is unfortunate, because developing graphicacy skills and visuospatial intelligence are particularly important in today's graphics-intensive, technologically driven world. *A picture is worth a thousand words* is oft-repeated for a reason, and Balchin and Coleman (1965:25) make the argument beautifully when they say that " . . . neither words nor numbers nor diagrams are simpler or more complex, superior or inferior. They are only more suitable or less suitable for particular purposes."

Maps and Graphicacy

In the world of GIScience, Jack Dangermond, founder of ESRI (Environmental Systems Research Institute), which produces *ArcGIS*, has declared the map to be ". . . one of humanity's most sophisticated conceptual creations" (Kimerling, et al., 2009, foreword). Although not always recognized as such at a quick glance, a map is a very complex graphic format. More than a photograph, the map is a purposefully simplified picture of the world; it is created to highlight certain aspects of reality, to provide an alternative that is less complicated, so the users may more easily focus on the highlighted spatial relationships the cartographer wishes them to consider. Compare, for example, two maps of Pakistan prepared by the Central Intelligence Agency. The physiographic map (Figure 2.3) is designed to highlight the physical features of the country: the river systems, the mountain chains, and their relationships to the major population centers, historical and archaeological sites. In the design of the transportation map (Figure 2.4), on the other hand, the cartographer has opted to remove the relief and the mountain chain designations, as well as the historical and archaeological sites. Instead, railroads, airports, canals, and roads have been emphasized, as well as their relationships to both major and minor population centers to highlight interconnectivity. Maps such as these are also distinctive because they offer us a view of the world that is outside our normal range of view, allowing us to visualize and understand spatial relationships over distances that are larger than we can see.

While the beauty of the map is its simplicity in communicating key spatial relationships, that built-in simplicity can be deceptive if the map's design is poorly executed, or if the user lacks knowledge that would increase his or her likelihood of misinterpreting the presented reality. Think about it: how many ways might there be to present a less complicated view of your environment? The answer can be unnerving, not only in *what* to show but in *how* to show it. As Harvey (2008) points out, this fuzziness in what to represent – as well as how to best represent it – leaves many feeling that these simplified representations are too complex, with the result that maps are easily misunderstood or perhaps even perceived as just too difficult to learn how to interpret properly. Like Harvey, however, I agree that taking the time to familiarize yourself with the principles and conventions of cartography and GIScience will significantly sharpen your visuospatial intelligence. It will also provide you with the information you

FIGURE 2.3
Pakistan's physiography.
Map: *Central Intelligence Agency.*

publications/map-downloads/Pakistan_Physiography.jpg

FIGURE 2.4
Transportation geography of Pakistan.
Map: *Central Intelligence Agency.*

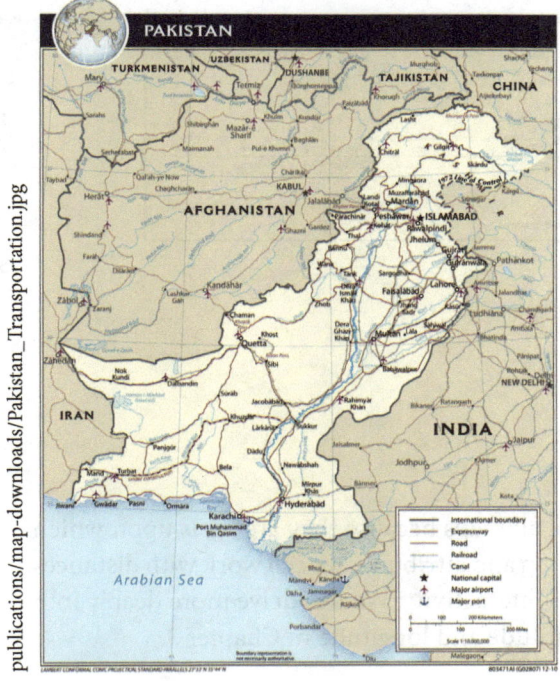

publications/map-downloads/Pakistan_Transportation.jpg

need to understand geospatial information, GIScience, and cartographic products, all of which are an increasingly important part of our everyday experiences.

Key Map Characteristics

Our fascination with maps has been part of our makeup since the ancient Greeks began investing time and energy in studying the size and shape of our planet, developing better methods of locating features, and plotting those locations on two-dimensional surfaces. As those methods have evolved, so has the variety of maps that can be produced, and the end products are often so different that there are times when one might question whether they all stem from the same set of principles. Luckily, they do, and one place we can start on our journey in exploring the underpinnings of GIScience is with those principles. In one sense, what we are broadly doing is defining a prototype for a map. The prototype will have all of the following six characteristics, as noted in Robinson, et al. 1995:

1. SPATIAL STRUCTURE

Our definition for *map* is quite broad and one of the reasons it works so well is because it highlights the idea of *spatial structure*, which emphasizes *location* and *attributes*, two key map characteristics we first introduced in Chapter 1 (Figure 2.5). *Location* is a place characteristic defined by using coordinate systems to designate a place's position on Earth. *Attributes* represent a name, label, measurement, or value associated with a location or pair of coordinates.

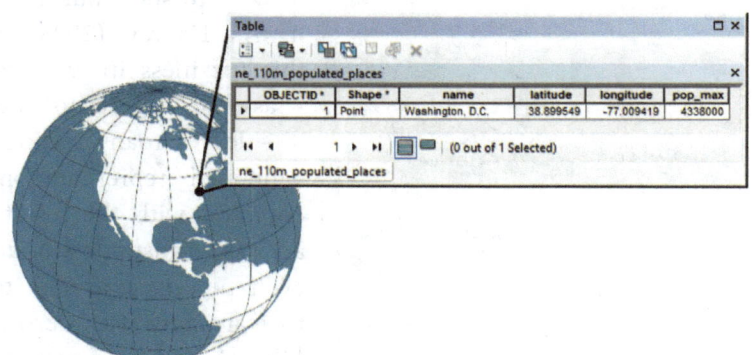

FIGURE 2.5
Spatial structure: Locations and attributes. Data: *Made with Natural Earth. Free vector and raster map data @ naturalearthdata.com.* Source: Elisabeth Nelson

These two characteristics are required of all maps because they are basis upon which we form spatial relationships. With location and attributes, we can work with distances and directions and evaluate patterns and interactivity. We will delve more deeply into location and the coordinate system of latitude and longitude in Chapter 3.

2. PROJECTION

Representing geographic information in two dimensions means that we must mathematically transform it from its three-dimensional source, the Earth. The systematic translation of locations on Earth's three-dimensional surface to their corresponding locations on a flat surface is called a *map projection*. Moving from a spherical coordinate system to one in which space is visualized using Euclidean geometry creates distortions in the spatial relationships between features being mapped (Figure 2.6). Map projections, then, play a vital role in map development. The projection that you choose to use will have consequences for your map's accuracy and the tasks for which it is designed. We will be learning more about map projections and their role in GIScience in Chapter 4.

FIGURE 2.6
The concept of a map projection. Data: *Made with Natural Earth. Free vector and raster map data @ naturalearthdata. com.*
Source: Elisabeth Nelson

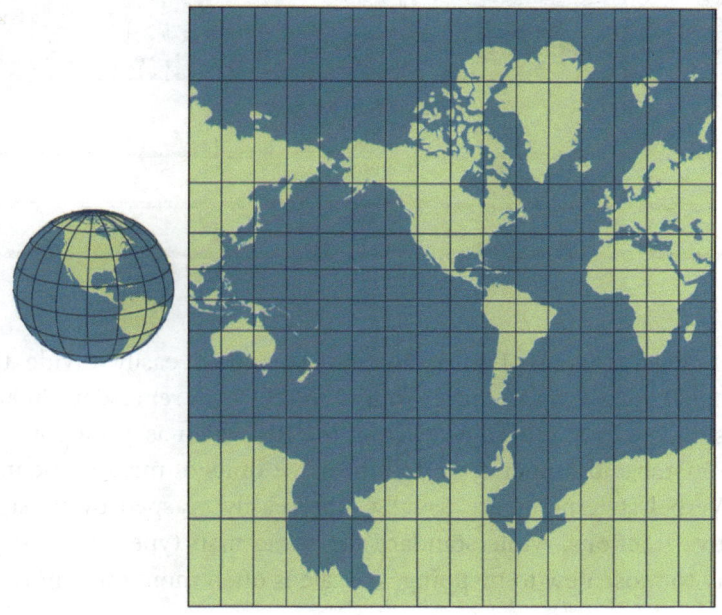

3. SCALE

Maps are not reality, but spatial surrogates; thus, they have a definite proportional relationship to the real world they represent. The relationship between the size of the area on the map and the size of the area in reality is described by the *map scale*. Mathematically, a map's scale is a *ratio* and is defined as *Map Distance/Ground Distance*. This ratio, when Map Distance is represented by a measurement of 1, is called a *representative fraction* (RF), and it is one of three ways of describing scale on a map. Take, as an example, a map that has a scale of 1/24,000 (or 1:24,000). On this map, 1 inch on the map represents 24,000 inches on the ground. In the case of Figure 2.7, the length of the road segment between Dogwood Drive and Starmount Drive, on the map, is approximately 1 inch. Since the scale is listed as 1:24,000, we know that the real-world distance is 24,000 inches.

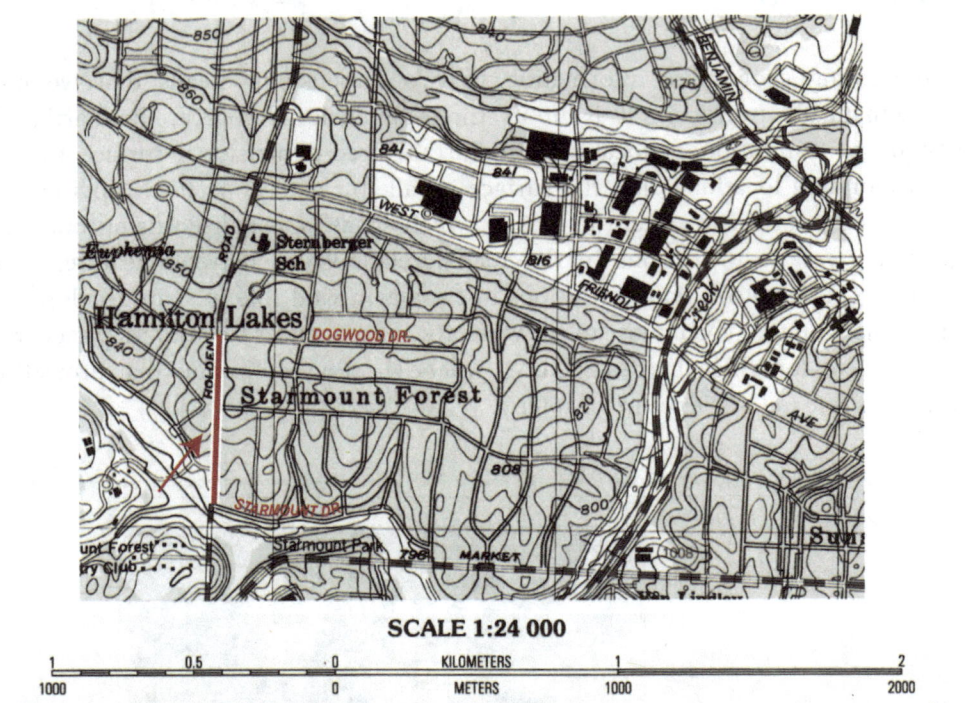

FIGURE 2.7
Map scale and the
representative fraction.
Map: *U.S. Geological
Survey, Department of
the Interior.*
Source: Elisabeth Nelson

SCALE 1:24 000

Using the English system of measurement, we could easily divide this ground distance (24,000 inches) by 12 inches to find that "1 inch represents 2000 feet." With this conversion, we have moved away from the RF, which as a fraction requires that both denominator and numerator be in the same units of measurement, to a *verbal statement*. Verbal statements are usually more easily grasped by most map users. Representative fractions, while standard for some map types, are not particularly user-friendly to those new to mapping, as scale is often more difficult to visualize in that form.

RFs are useful, however, because they are not tied to any particular unit of measurement. By keeping both sides of the fraction in the same units, any system of measurement may be used. Thus, the 1/24,000 could mean, "1 inch represents 24,000 inches," "1 foot represents 24,000 feet," or even "1 meter represents 24,000 meters." Imagine the following: you have an image on which the length of an American football field is approximately 1 inch (Figure 2.8). The actual ground distance of that field is approximately 360 feet, or 4,320 inches (360 ft. x 12 in.). The representative fraction of the image, then, according to our definition, is 1/4,320, which may also be written as 1:4,320.

FIGURE 2.8
Setting up the
representative fraction.
Image: *U.S. Geological
Survey, Department of
the Interior.*
Source: Elisabeth Nelson

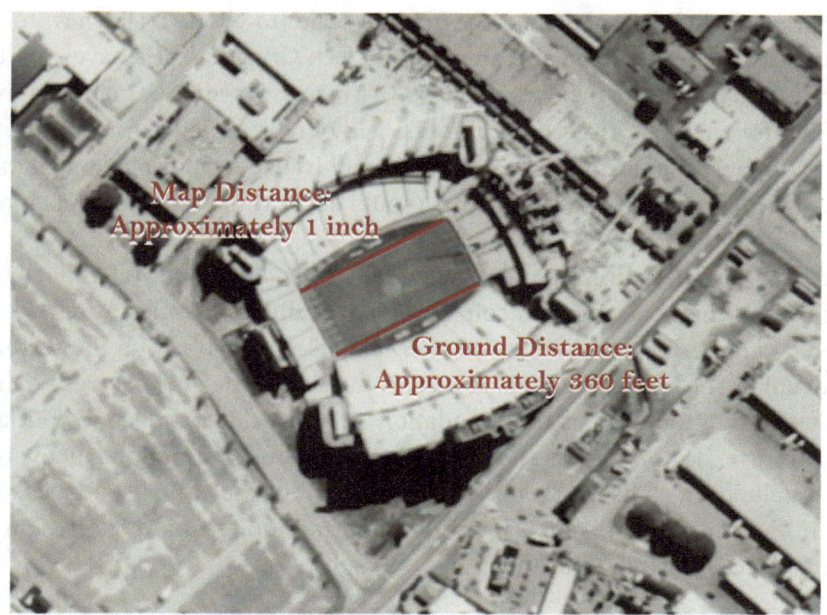

If you were to measure that same football field in centimeters, it would be 2.54 cm in length. The actual ground distance, using the metric system, is 109.73 meters or 10,973 cm (109.73 m x 100 cm). That leaves us with a fraction of 2.54/10,973. To find the RF, the numerator must be 1, which requires us to divide both the numerator and denominator of the fraction by the numerator value (2.54 in this case). When you do that, the resulting RF is 1/4,320.

Another more user-friendly format for scale is the *graphic scale*. A graphic scale is simply a line on the map that is subdivided into major units representing ground distances. In Figure 2.9a, a line 1 inch long has been drawn to represent map distance. It is labeled with the corresponding ground distances for our football field example, as specified by the RF and/or verbal statement. By measuring the length of the line, and associating that length with the ground distance, one may easily convert it back to a word statement or RF (in this case, 1 inch represents 360 feet).

FIGURE 2.9
Examples of scale
designs for a map with
a scale of 1:4,320.
(a) 1-inch scale length,
and (b) modified scale
lengths to show rounded
ground distances
Source: Elisabeth Nelson

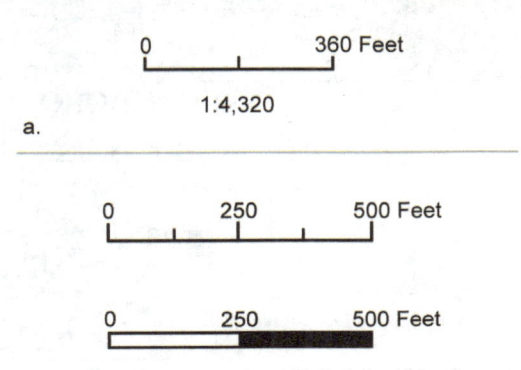

In Figure 2.9b, you can see that both the length of the graphic scale and its overall design may vary; these are decisions made by the producer of the map. In this instance, the cartographer slightly lengthened the graphic scale (to about 1.4 inches) to round the ground distance it represents to an even 500 feet. Graphic scales have a particular advantage in that they retain their accuracy even if a map user prints the map then enlarges or reduces it for later use. The same cannot be said of representative fractions or verbal statements; because they are numerical and not graphical, reducing or enlarging the map does not reflect the change in numbers that should occur with these manipulations.

Map scale may also be expressed more generically as a *relative scale* (Campbell, 1998), which is useful when one wants to compare maps at different scales. A relative scale, expressed as *large-scale, small-scale,* or *intermediate-scale,* gives one the feel for the level of detail expressed on the map, as well as the size of the area covered. Using the representative fraction (RF) as the foundation, a large-scale map is one with a large RF. This, however, is not quite as simple as it sounds. Think of fractions that you work with on a daily basis: ½ is a larger fraction than ¼, but the scale itself has a *smaller* number in the denominator; this fact of life can cause quite a bit of confusion when working with scale cartographically.

There are also no hard and fast rules on what constitutes a large- versus a small-scale map; scale, in fact, is best thought of as a continuum (Figure 2.10). That said, generally, a large-scale map is 1:50,000 or larger; it covers a limited area, but is able to provide a lot of detail about the place being mapped (Robinson, et al., 1995). A small-scale map, on the other hand, is typically 1:500,000 or smaller; it covers extended areas and provides minimal detail about the mapped features. Those in-between are labeled medium-scale. Because map scale directly affects the level of detail that can be represented on a map, this concept, along with map projection, is another important key to understanding map accuracy.

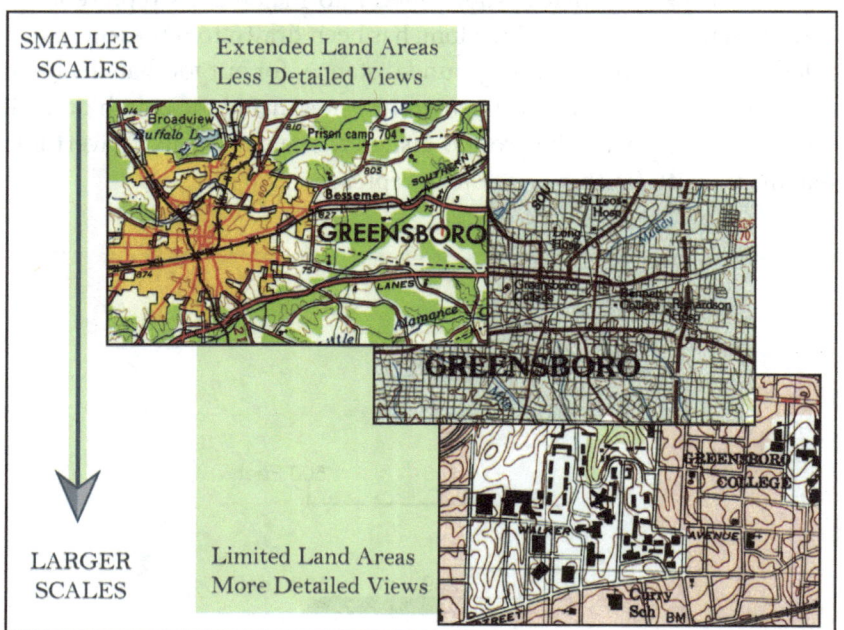

FIGURE 2.10
Scale as a relative concept.
Maps: *U.S. Geological Survey, Department of the Interior.*
Source: Elisabeth Nelson

4. GENERALIZATION

Since every map is, by necessity, smaller than the real world it represents, there is no way to show every detail of reality, even on a large-scale map. This basic fact about maps means that the level of detail provided on any given map must be manipulated to suit its purpose. Think of it like this – on a map of 1:65,000, you want to show a park that is approximately 1 mile along its widest length in the real world. How big should that park be on your map? According to the RF, 1 inch on the map represents 65,000 inches on Earth. If we divide 65,000/12, we know that 1 inch on this map represents 5,417 feet (or slightly longer than a mile) on the longest side (Figure 2.11a), which would make the park a little less than 1 inch along its longest side.

That's quite a reduction from the real-world size, but let's not stop there. Let's see what we would have to do to symbolize the park on a map at a scale of 1:650,000. Our RF here stipulates that 1 inch on this map represents 650,000 inches on Earth. If we divide the 650,000 inches by 12, you have a verbal statement that says that 1 inch represents 54,167 feet. On this map, then, 1 inch represents much more than 1 mile, so we can no longer draw the park as a 1 inch in length. If we divide 54,167 feet/5,280 feet, we see that 1 inch on this map actually represents about 10.3 miles. To fit the park on this map, then, at a size commensurate with the scale, the long side of the park would have to be about 1/10 the length on the previous map, or about 0.1 inch in length. The 1:650,000 map is much smaller-scaled and can't support the same level of detail as the 1:65,000. How, then, should we represent the park on this new map? One possibility would be to change the symbology from an area symbol to a point symbol (Figure 2.11).

This is the essence of *cartographic generalization*, which is a suite of processes cartographers use to select which information to retain for a given map and to determine how to represent that information within the context of the final map scale.

Many GIScientists have studied cartographic generalization, and several have proposed formal models of the process to help cartographers understand its implications for map creation and map use. Of all the operations that are part of this process, there seem to be a core few that everyone includes and that you should be aware of as you explore the world of GIScience. These are *selection,*

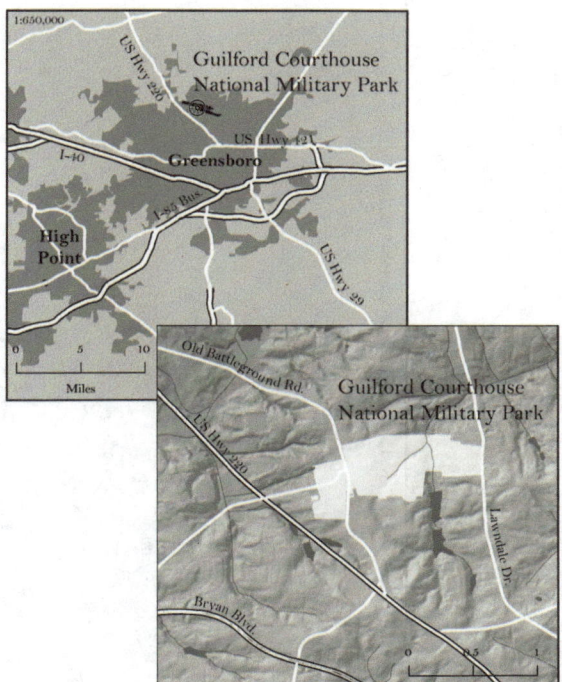

FIGURE 2.11
Examples of the influence of scale on map design: (bottom) map of Guilford County National Military Park at 1:65,000; (top) map reduced to 1:650,000 with change of symbology to accommodate smaller scale.
Data: *U.S. Census Bureau; National Park Service.*
Source: Elisabeth Nelson

simplification, symbolization, and *classification* (Dent, et al., 2009). Implementing each of these processes results in a reduction of the amount of detail included on a map. How much reduction is largely a function of the map's scale, with small-scale maps requiring more generalization due to space limitations than their larger-scale counterparts do. While generalization provides a simpler picture of reality, if done correctly it will retain the distinguishing characteristics of mapped features and leave the user with enough information to grasp the conceptual meaning of the map.

Selection. Cartographic generalization begins with the process of *selection*; in fact, Robinson and colleagues (1995) see selection as a precursor to other geometric and statistical generalization operations. Selection is simply choosing which features to show and which to eliminate. The choices that are made depend on the purpose of the map and the final scale, i.e., how much room there is to show detail. Consider, for example, the difference between a road map for the state of Colorado versus one focusing on the Denver, Colorado region (Figure 2.12). The scale of a map of Colorado, designed for a 2.75 x 2.125 inch space, would be around 1:9,000,000 (1 inch represents 142 miles). At this scale, you could show the major cities of the state, the interstate system and some of the other regionally important roads. A map of the Denver region of Colorado, however, using approximately the same page constraints, would result in a final map scale of approximately 1:2,000,000. Not only could you show the major cities, interstates and regionally important roads, but you could also display smaller cities and towns, minor roads with labels, and other labeled physical and cultural features, such as rivers, streams, parks, forests, historic sites, and outlines of urbanized areas.

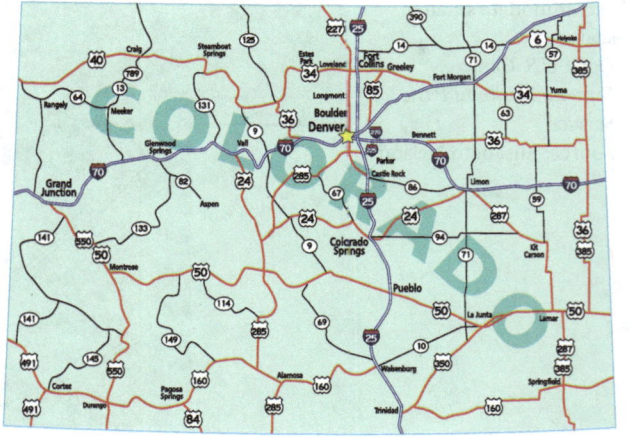

Image © Globe Turner, 2013. Used under license from Shutterstock, Inc.

© Stacey Lynn Payne/Shutterstock.com

FIGURE 2.12
The process of selection in cartographic generalization: (left) map of the Denver, Colorado region at 1:2,000,000; (right) map of Colorado at 1:9,000,000.

Simplification. You may also find that some features need to be *simplified* at smaller scales. In the case of a United States reference map, for example, that might mean using less complex lines to represent coastal boundaries; these lines would retain the major characteristics of the coastline, but eliminate or smooth out fine details so that the final appearance is simplified and less cluttered. In Figure 2.13, the 1:50,000,000 database would represent such a simplified coastline. This database, however, would not be appropriate for mapping a much larger-scaled map, such as a portion of the North Carolina coastline. You would not choose to simplify shorelines to the same extent at this larger scale because the final appearance would look coarse and result in significant map errors and inaccuracies.

FIGURE 2.13

Simplification and the relationship of scale to GIS database choices. Data: *Made with Natural Earth. Free vector and raster map data @ naturalearthdata.com;* National Weather Service. Source: Elisabeth Nelson

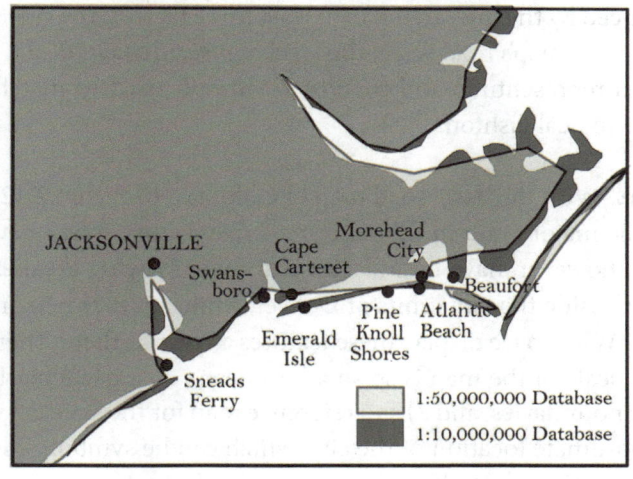

To better visualize these potential problems, compare the coastal linework of the two database scales. The coarseness of the 1:50,000,000 database means the some of the smaller towns and cities on the coast, such Beaufort, Swansboro, and Sneads Ferry, are located in the water. The cities and towns database was compiled at a larger scale than the 1:50,000,000 shoreline database, introducing error into a map comprised of these two layers. A much better choice would be to use the 1:10,000,000 database with the cities and towns data, as this pairing of data layers provides much more detail in the shape of the coastline and better matches the scale of the cities and towns database. It should be clear from these examples that, despite the tendency one has to treat GIS data as *scaleless*, that is most definitely not the case. You can zoom in and out on a GIS database, but that does not mean that the level of detail of the database changes. This example also serves to point out that today's GIS databases are very similar to traditional map output; that is, they are compiled (often from paper maps) at a specific scale, which limits them in their mapping applications to a range of scales around that original compilation scale. Similar to "buyers beware", there should also be a "mappers beware" notice to that effect; if the database scale is unsuitable for your needs, you need to find a different, more appropriately scaled database before continuing with your mapping project.

5. SYMBOLIZATION

As the ability to include more detail increases, then, we can often choose symbology that looks more like the real-world feature. This can be seen with the coastlines in Figure 2.13 and with the change in symbolization for the park at different scales in Figure 2.11. Parks after all, aren't really points; they all have boundaries and thus, areal extents. It just isn't always convenient – or possible – to show that symbolically on a map, which leaves us with a more abstract and generalized symbol choice. Because of this added complexity, *symbolization* has a whole host of principles and conventions associated with it, so much so that it often is addressed not only as *part* of **generalization**, but also as a basic characteristic of maps as well. What are some of these principles and conventions? Far from a random process, the selection of map symbols is influenced by the interaction of at least three factors: the *symbol dimension*, which is related to the map's final scale; the *level of measurement* of the attribute data that the symbol is representing; and the *visual variable* used to match that level of measurement in a logical fashion.

Symbol Dimension. With the basic road map of Colorado in Figure 2.12, for instance, the locations of the major cities in the state are displayed using point symbols. Cities in the real world, however, have boundaries and cover varying areal extents so that someone may live inside the city limits of Denver, while another person lives outside those boundaries. Why do we display these features as points, then? There are a couple of reasons: 1) the scale of the map is so small it doesn't do a particularly good job of showcasing those boundaries, and 2) as a reference map for the state, all we really need to know is the coordinate location of the city, which can be symbolized using a point symbol. So, although the city itself has areal extent and may be shown as a polygon in its true form (see Denver in Figure 2.12), at smaller scales and for particular purposes, cartographers may choose to show it using a point symbol, which has location, but no extent. This is the concept of *symbol dimensionality*. All mapped features have a real-life dimensionality and a symbolized dimensionality, but the mapping from reality to symbol is not necessarily one-to-one, as we just saw. In most situations, you will work with either *points* (0-dimensions), *lines* (1-dimension with a length attribute), or *polygons* (2-dimensions with length and width attributes).

Visual Variable. The choice of symbol dimensionality is made in concert with the choice of which *visual variable* to use in drawing the symbol. The term "visual variable" is associated with Jacques Bertin (1967), a French cartographer who used it to describe the characteristics of a symbol that might be harnessed to help map users perceive spatial variation in attributes across the mapped features. There are seven traditional visual variables that Bertin associated with map symbols: *position, shape, size, hue, value, orientation,* and *texture.* Other researchers have since supplemented those, but these will suffice for our explorations here. As Figure 2.14 shows, *position* simply denotes coordinate locations, in this case for cities, which have been symbolized here by using black dots, all of which are the same size.

FIGURE 2.14

Bertin's original visual
variables.
Data: *Made with Natural
Earth. Free vector
and raster map data
@ naturalearthdata.
com; National Weather
Service.*
Source: Elisabeth Nelson

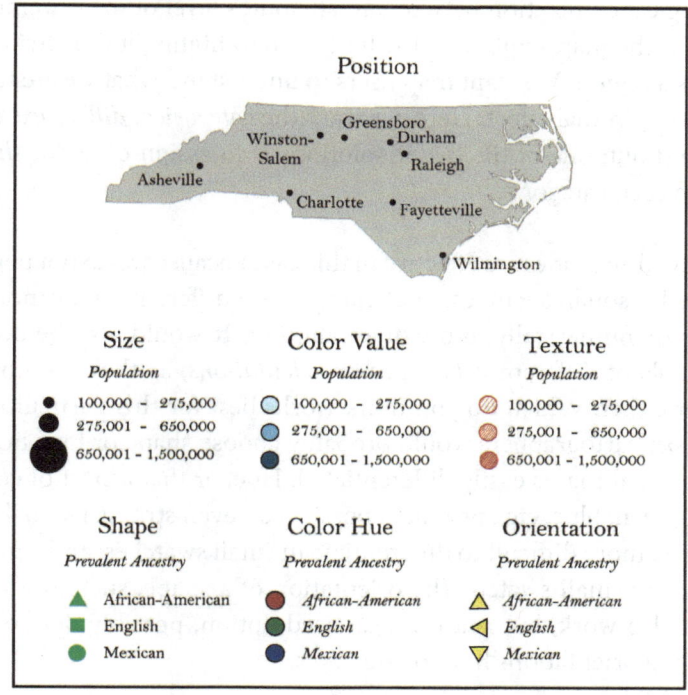

What if, however, we wanted indicate the prevalent ancestry of residents for each city? From Bertin's perspective, you could implement this in a variety of ways. For example, you might symbolize those cities reporting a prevalent African-American ancestry with triangles, whereas you could use squares for those reporting a prevalent English ancestry. This type of choice takes advantage of the visual variable *shape*. Alternatively, you might symbolize each city with a circle, and then vary *color hue* of the circles (i.e., red, green, blue) to show that distinction. Would symbolizing prevalent ancestry by varying the *sizes* of the circles for each group be a more effective or less effective solution? The answer to this type of question lies in the attribute data's *level of measurement.*

Levels of Measurement. Attributes may be described using many scales of measurement; they are often classified by both formal mathematical properties and by the operations we use during actual measuring (Stevens, 1946). Particularly important to GIScience is that the way in which an attribute has been measured and recorded directly affects how we may manipulate it statistically and visually. For instance, in the context of levels of measurement, four measures can be used to manipulate attribute symbolization: *nominal, ordinal, interval,* or *ratio.*

An attribute with a *nominal* level of measurement is one that only allows us to determine whether two or more categories are alike or different. In a case like Figure 2.14, we would have nothing more than a label, such as those describing prevalent ancestry for each city. Prevalent ancestry is not a descriptor that can be ordered; one type of ancestry is not better than another and they do not differ numerically, they are simply different in kind or category. In mapping, one of the key goals is to create an intuitive and logical connection between an attribute's level of measurement and the visual variable of the map symbol that will be used to highlight the attribute's spatial variation across a region. We want map users to understand what we are trying to say quickly and with minimal effort. Here, we want the *categorical difference* of prevalent ancestry to stand out; one of the better solutions is to assign different *shapes* of the point symbol to each category.

Differences in symbol shapes are effective in this case because we do not perceive them as being ordered – squares, circles, and triangles are different, one is not associated as being better or numerically 'more' than another. It would also be acceptable to use point symbols of different *color hues* or *orientations*, as they are not perceived as having order either. Which do you think works best for this particular dataset in Figure 2.14? Most cartographers would probably choose shape or hue; both produce a strong visual signal that is easily differentiated. Hue, or that aspect of color that we identify as red, green, blue, etc., produces possibly an even stronger visual signal than shape, but may be more difficult to differentiate in small swatches, and a point symbol is typically a pretty small swatch. The orientation of a shape, such as a rectangle or triangle, might also work, but is a less exercised option, possibly due to the jarring effects the varying orientations have on our eyes.

If our dataset recorded not a label, but a number for each city's total population, also shown in Figure 2.14, then you would have a *ratio* level attribute measurement. Ratio levels of measurement are associated with numbers; there is an explicit ordering of the cities on the map according to population size in this case, so there should be a corresponding order to the visualization of the attribute spatially. A classic option here would be to vary the *size* of the original circle symbol. Smaller to larger circles have a definite implied order with which all map users can identify. *Color value*, the relative lightness or darkness of a single hue, would also work. However, the size of the point symbol produces the same problem for value as it does for hue – it may not have enough of a canvas for map users to differentiate differences easily. *Texture*, the spacing of elements within a pattern, is also an ordered visual variable, but it typically does not work well with point symbols.

I should point out here that there is a hierarchy implied in the concept of 'levels of measurement' (Figure 2.15). Nominal data are characterized by the least amount of detail and ratio data the most. The next level above nominal is *ordinal*. Ordinal measures are *rankings;* an example might be a ranking of each city's unemployed, such as below the national average, average, and above the national average. There is an order to the attribute that can be visualized symbolically, but the distance between the categories cannot be described numerically, just as more or less. Levels of measurement not attached to numbers, such as nominal and ordinal, are more broadly categorized as *qualitative data*. With ordinal level data, you not only have a label, but a descriptive indication of ordering. In this case, you would most likely want to emphasize the difference in that rank-order on your map, which means using a visual variable that map users intuitively 'see' as having order. Despite the fact that no numbers are associated with ordinal data, the data variation itself is displayed quite similarly to ratio and interval level data and makes use of the same visual variables.

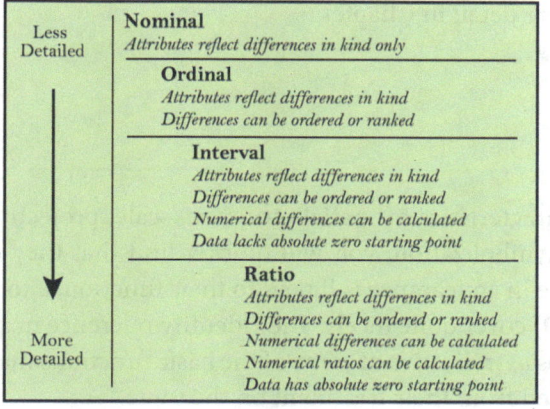

As you move from less detailed data to more detailed data, each specific level always include all the qualities of the less detailed levels that come before it. The next step up from ordinal level data adds meaning to the numerical differences between categories. Both *interval* and *ratio* level attributes are examples of *quantitative data* – they are based upon numerical differences. In our example, if our dataset included attributes that listed the number of unemployed for each city, then we would be able to calculate the precise difference in unemployed between cities. This increased level of detail provides additional interpretative power to our visualizations.

I have lumped interval and ratio data together in the previous paragraph, but there is a key statistical difference between them, even if there is not necessarily a graphical difference to how we visualize the data. In our example, we moved from nominal to ordinal to ratio data, skipping interval. Why? When talking about the number of unemployed, there is an absolute zero involved; in other words, you can create meaningful fractions or ratios from the data. It would be, for instance, quite proper to state that one city has approximately 1.7 times as many unemployed as another. This is something that cannot be done with interval data, which does not have an absolute zero. Interval data are numerical; it just doesn't make sense to calculate fractions or ratios using it. The classic example of interval data is the Fahrenheit temperature scale. With temperatures, the intervals between the values are interpretable (there is a change of 20 degrees between 20-40 and 40-60). However, because zero has no absolute meaning in this system – no set number denotes a lack of temperature, you cannot say that 40 degrees is twice as hot as 20 degrees.

6. CLASSIFICATION

Classification refers to the process of grouping similar attribute values. Like symbolization, there are several principles and conventions associated with the process, making this not only an operation within the context of cartographic generalization, but also a basic map characteristic. The goal of grouping similar attribute values is to reduce the complexity of the final map; thus, instead of trying to display each numeric value in the dataset with a unique symbol, the cartographer uses a smaller set of symbols to display a limited number of groups that house similar values. The result, if conducted with care, improves the organization of information, making it easier for the map user to visualize the broader spatial patterns within the data. In Figure 2.14, for example, the major cities in North Carolina could be grouped or classified according to population size; in this case, cities with larger populations might be represented with larger symbol sizes than cities with smaller populations. We will explore classification in more detail in Chapter 6.

Functions of Maps

While all maps share the basic characteristics of spatial structure, scale, projection, generalization, classification, and symbolization, you will quickly find that they do **not** all look the same. The difference in appearance is linked to their function, a topic that Kimerling and colleagues (2009) cover quite nicely. They identify reference maps, thematic maps, and navigation maps as indicative of some of the basic functions maps serve. The following is largely derived from their treatment on the topic.

Reference maps are designed so map users can identify the locations of several different feature types quickly and easily, as well as determine the distances and directions between them. These maps tend to be larger in scale, and to include as much detail as possible and still be accurate. Another trademark design of these maps is their attempt to portray all mapped features as equally important from a visual perspective. The classic example of a general reference map is the topographic map, which is designed to highlight the physical and cultural characteristics of an area (Figure 2.16). In the United States, the United States Geological Survey (USGS) produces our topographical map series. These maps, which we will discuss in detail in Chapter 5, not only provide the map user with locations of such features as roads, shorelines, water features, and buildings, they also use specialized symbology to embed the shape of the land into the map.

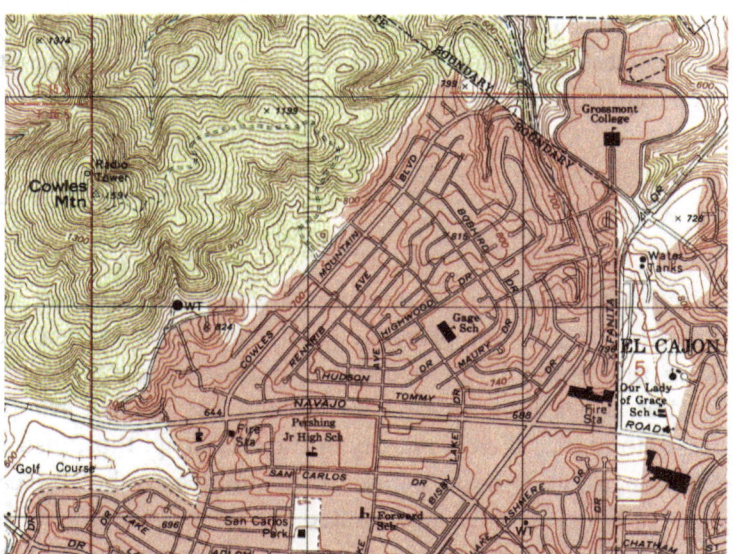

FIGURE 2.16
Example of part of a 7.5 minute USGS topographic sheet. Map: *U.S. Geological Survey, Department of the Interior.*

Similar to reference maps, but designed for hands-on calculations, are navigation maps or charts. These are maps designed for navigators, who use them to plot courses and mark bearings (the topographic map also fits here, so you can see these functions are not exclusive of one another, but rather form a continuum of uses). They include nautical charts and aeronautical charts among others. Nautical charts are designed for water navigation, and thus, will have detailed shorelines and detailed water depths available to the navigator. Air navigators use the aeronautical chart, which shows such attributes as heights of towers, possible obstructions, and specialized symbology for ground elevations. You could also include the standard road map in this group, although its 'look' is a bit different from the chart. Nevertheless, a good road map supplies information necessary to route planning: road quality, distances, and places of interest.

Thematic maps, on the other hand, function quite differently (Figure 2.17). These maps have been likened to geographic essays in that they focus on one particular topic. The topic can be physical in nature, such as the distribution of soil types in N.C., or they can choose to emphasize a more abstract, cultural topic, such as the varying chances of contracting malaria worldwide. Because they visually emphasize one particular attribute over the rest of the geographical background, these maps communicate quite different information. Typically smaller in scale, they are not suited for estimating distances and directions or locating specific geographic features. Instead, thematic maps are built specifically to communicate information regarding the spatial structure of the topic emphasized.

FIGURE 2.17
Example of a thematic map.

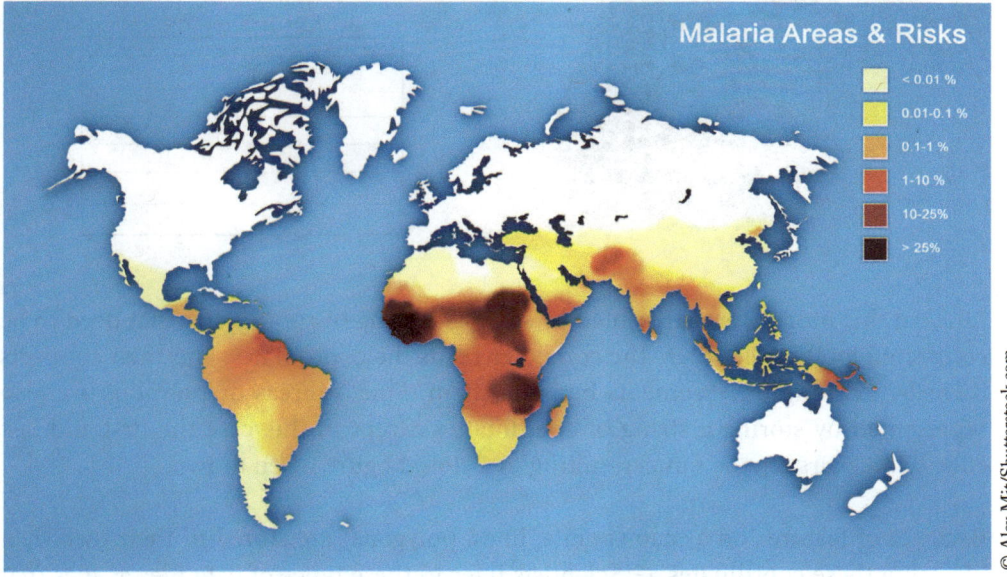

Maps in the Digital Environment

Moving from a real-world, printed map to a computer representation is known as *data modeling*. On the surface, this might not seem difficult – until you realize that computers can't 'see' graphics and images like we can; it deals only with numbers. One of the first challenges you confront then, is how to model our world so the computer can 'see' what we as humans see and interpret when a map is put in front of us. How do you convert lines, shapes, and images into numbers? One solution is to use geometric shapes, such as points, lines, and areas, to represent different map features (Figure 2.18). This is called a *vector* model of reality, and the point – represented by an (x,y) coordinate tied to a spatial reference system (such as latitude and longitude) – is the simplest geographic feature. It could be a city, or an archaeological site, perhaps. Storing these features as (x,y) coordinates allows us to do at least three things: 1) transfer the data to the computer, 2) use the data to redraw the map digitally, and 3) perform some very basic calculations, such as the distance between two features.

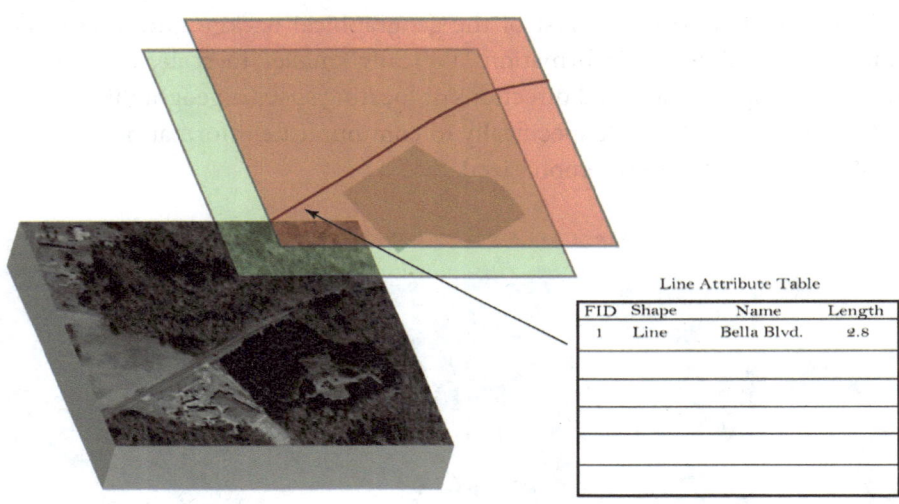

Line Attribute Table

FID	Shape	Name	Length
1	Line	Bella Blvd.	2.8

FIGURE 2.18
The world as seen in a vector model – points, lines, polygons. Photo: *U.S. Geological Survey, Department of the Interior.* Source: Elisabeth Nelson

Lines in this model are represented by storing a string of coordinates that need to be connected by the computer. The computer reads the coordinates, and then connects them by drawing line segments between them. Finally, areas or polygons are also represented by storing a string of coordinates, except in this case the first and last coordinate pairs are the same, resulting in a closed figure when drawn.

Each set of features on a map (points, lines, polygons) are stored in their own data file using these coordinates. GIScientists refer to these types of data files as files that contain "feature geometry." Stored in this manner, the computer can now draw the features, but it still has no way to differentiate them symbolically within any data file. For this to occur, we need to supply the computer with *attribute data* for each feature. Attribute data are recorded in a flat file, or spreadsheet form, for each geographic feature. This attribute file is linked to the file of feature geometry using a Feature ID

Number (FID). The FID is included in both the geometry and attribute files, and is a unique code supplied to each geographic feature (Figure 2.18). Because it exists in both files, the computer can use it to associate the correct attribute data with the correct geometry. For example, by linking city coordinates (points) to a table that lists the population for each city, we can use different circle sizes for each city, depending on how many people reside there. This setup, where we use multiple tables that are linked by FIDs, is called a *georelational database*.

References

Aldrich, F., and L. Sheppard. 2000. Graphicacy: the fourth 'R'? *Primary Science Review*, v. 64: 8-11.

Balchin, W. G. V., and A. M. Coleman, 1965. Graphicacy should be the fourth ace in the pack. *The Times Educational Supplement.* (Rpt. in *The Cartographer,* 1966, 3 (1), 23-28).

Bertin, J. 1983. *Semiology of Graphics: Diagrams, Networks, Maps.* Madison, Wisconsin: University of Wisconsin Press.

Dent, B.D., Torguson, J.S., and T.W. Hodler. 2009. *Cartography: Thematic Map Design.* 6th ed. New York, New York: McGraw-Hill.

Harvey, F. 2008. *A Primer of GIS.* New York, New York: The Guilford Press.

Kimerling, A.J., Buckley, A.R., Muehrcke, P.C., and J.O. Muehrcke. 2009. *Map Use: Reading and Analysis.* 6th ed., Redlands, California: ESRI Press Academic.

Poracsky, J., Young, E., and J.P. Patton. 1999. The Emergence of Graphicacy. *The Journal of General Education*, v. 48(2): 103-110.

Robinson, A.H., Morrison, J.L., Muehrcke, P.C., Kimerling, A.J., and S.C. Guptill. 1995. *Elements of Cartography.* 6th ed. New York, New York: John Wiley & Sons, Inc.

Exercise 2
Visualizing Spatial Features in ArcMap

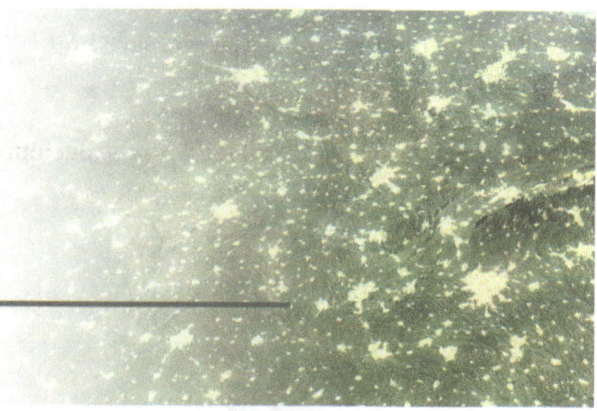

In Exercise 2, you will create a general reference map of South Carolina and use it to explore the map characteristics of location/attributes, map projection, map scale, and the processes of cartographic generalization.

OBJECTIVES

- Create an **ArcMap** project
- Verify coordinate systems for data layers
- Choose appropriate data layers as the foundation for your map
- Set map scale and create scale representations
- Create effective map symbols

SOFTWARE INFORMATION

Introducing ArcMap uses ESRI's **ArcGIS 10.6** software. **ArcGIS** is a commercial GIS package, available in geospatial labs affiliated with schools that have a campus-wide site license for the software. Instructors at these campuses may also request 1-year student versions of the software at http://www.esri.com/landing-pages/education-promo.

DATA

The data for the exercises is available from the Kendall Hunt Student Ancillary site. See the inside front cover for access information. You may also be directed to download the data from a different location by your instructor.

1. Download the data as instructed by your instructor
2. Save the file to a location where you have read/write privileges (USB key, home computer, class server)

The file you just saved is in compressed (*.zip) format, and was created using **7-zip** freeware (http://www.7-zip.org). To use the data for this exercise, you must decompress it using the same software.

3. Locate the zipped file that you saved
4. Right-click the file
5. Choose *7-Zip - Extract Files*
6. Click *OK* to create a folder with the decompressed data

Creating an ArcMap Project

CONCEPT

1. Start **ArcMap**. You can access ArcMap using several different methods, depending on your environment. You can start by clicking the Windows start menu icon at the bottom left corner of your screen, then navigate to the ArcMap icon in the ArcGIS folder, or search for ArcMap in the search window by typing "arcmap".
2. In the "*ArcMap - Getting Started*" dialog box, click *Cancel*
3. If the *Catalog* window is not open, click on the *Catalog* tab on the right side of the **ArcMap** interface OR click the *Catalog* icon on the *Standard* toolbar
4. Click the pushpin in the upper right corner of the *Catalog* window to freeze the window in place (Figure 2.1)

FIGURE 2.1
Used with permission.
Copyright © 2018
Esri, ArcGIS, ArcMap,
ArcCatalog, United
States Department
of the Census;
naturalearthdata.com
All rights reserved.

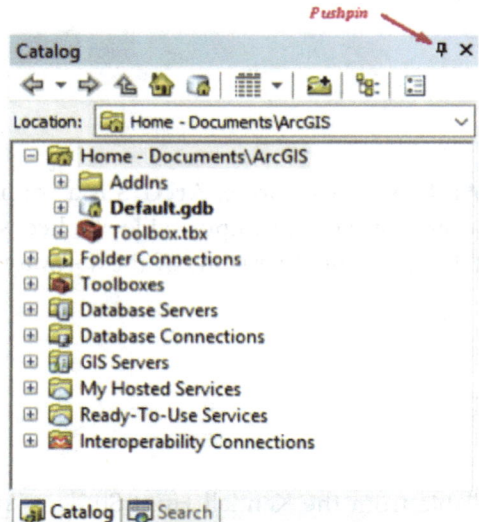

5. Right-click *Folder Connections* and choose *Connect to Folder*
6. Navigate to the location where you saved the *Chapter02* folder of data for this exercise
7. Click once to select the *Chapter02* folder, then click *OK* to create a folder connection that will allow you to access the data inside the folder

8. Under *Folder Connections*, click the + sign to the left of the folder to see its contents. You should see a geodatabase for South Carolina (*Exercise_02.gdb*)

9. Click the + sign to the left of the geodatabase to see the actual spatial data files you will use in the exercise

Verifying Coordinate Systems

CONCEPT

Representing geographic information in 2 dimensions means that we must mathematically transform it from its 3-dimensional source, the Earth. The 3-dimensional source of spatial data is called the *geographic coordinate system*. Its systematic translation of locations to their corresponding locations on a flat surface is called a *map projection* or *projected coordinate system*.

For the files or data layers in this geodatabase to display together accurately on the same map, they should each have the same coordinate systems. Verify this property for each file in the geodatabase by following these steps:

1. In the *Catalog* window, right-click each data layer, then click *Properties*
2. Click the *XY Coordinate System* tab

QUESTION 2.1

Of the 4 data layers in the South Carolina geodatabase, one has a coordinate system that does not match the rest. Which layer is it?

a. SC_Dtl_Counties
b. SC_Gen_Counties
c. SC_MjrCities
d. SC_MjrRds

3. When finished, click and drag each data layer from the *Catalog* window to your *Data View* window
4. If a Warning dialog pops us, click *Close* to move through it – we'll address this later

---

FIGURE 2.2

5. In the *Table of Contents* (TOC) window, click on the *List by Drawing Order* icon (Figure 2.2)
6. Click and drag the data layers as needed to re-order them within the *data frame* or map (called *Layers* in the TOC); SC_MjrCities should be listed first, followed by SC_MjrRds, SC_Dtl_Counties, then SC_Gen_Counties (Figure 2.3)

FIGURE 2.3

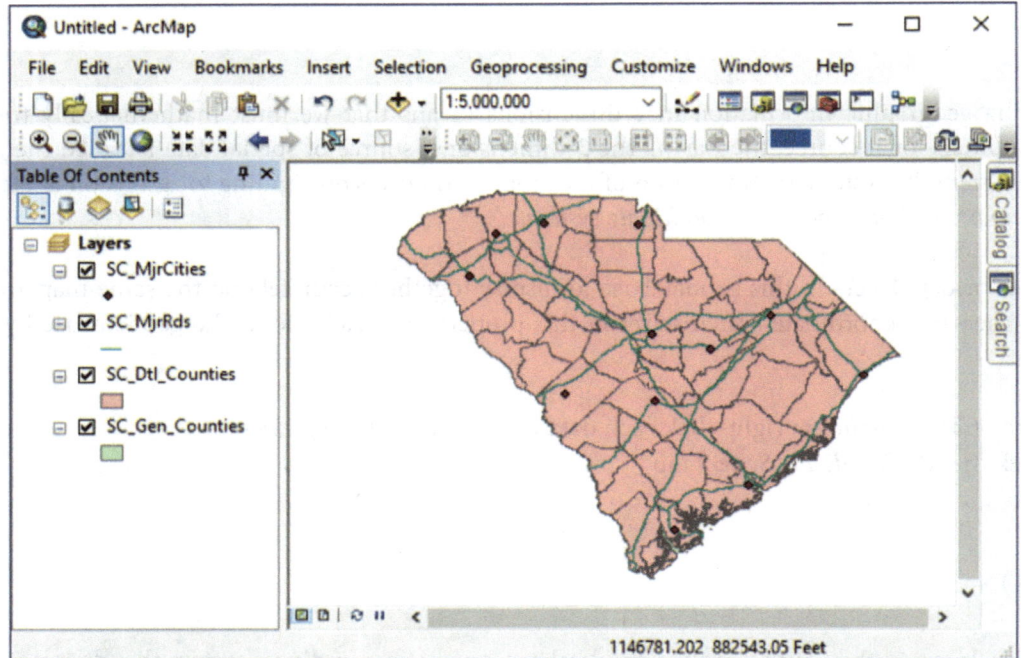

Saving a New Map Document

CONCEPT

Before saving a new map document for the first time, there are two settings you should check: the *default geodatabase* and *how your document references its spatial data*.

Each map document has a default geodatabase. This is the location where datasets you create via editing and geoprocessing operations are stored, so you will want to set it to the geodatabase your current map document is using.

1. From the *File* menu, choose *Map Document Properties*
2. Click the file folder button to the right of the default geodatabase path and browse to the *Exercise_02.gdb*
3. Click to select it, then click *Add*

It's also important to remember that your map document does not store the actual spatial data it displays—instead, it stores the *path* or *address* to where that data resides on the computer. When you open a map, if the path to a layer of data has changed (maybe because you moved the file to a different location), **ArcMap** will not be able to locate it and draw it for you. Choosing to store your document's paths as *relative pathnames* can lessen this problem by determining paths to data relative to the map document's current location in the file system rather than by using absolute paths requiring a drive letter.

4. Next to pathnames, place a checkmark by *Store relative pathnames to data sources*
5. Click *OK*
6. From *File*, choose *Save As*
7. Name your file using this convention: *lastname*_Exercise02
8. Save your map to the *Chapter02* folder containing *Exercise_02.gdb*
9. Double-click the *Computer* icon your desktop
10. Navigate to the location where you think you saved your file and verify that it is there

Choosing Data Layers for Map Creation

CONCEPT

Casual users of GIS have a tendency to treat GIS data as *scaleless*, because they can zoom in and out on the data layers. In reality, however, while zooming in or out may change the scale at which the database is viewed, it does not change the level of detail that is built into the data. Today's GIS data are still very similar to traditional map output; that is, they are compiled (often from paper maps) at a specific scale, which limits them in their mapping applications to a range of scales around the original *compilation scale*.

In this section, you will explore the *metadata* of each data layer to see how their scales relate to one another.

1. In the TOC window, right-click the *Layers* data frame, then select *Turn All Layers Off*
2. Turn on the *SC_Gen_Counties* layer
3. Using the **Data View's** *Zoom In* tool, zoom in on a section of the coast
4. Turn on the *SC_Dtl_Counties* layer
5. Zoom in on the coast some more to examine the differences between the two data layers
6. Check and uncheck the layer that is drawn on top to see the differences more clearly
7. From the *Customize* menu, choose *ArcMap Options*
8. Click on the *Metadata* tab
9. Under *Metadata Style*, choose *North American Profile of ISO19115 2003* and click *OK*
10. In the *Catalog* window, right-click each data layer in turn, then click *Item Description*. Scroll through the metadata until you find the *Resource Details* heading
11. Make a note of the *Spatial Resolution* or *Compilation Scale* of each data layer
12. Close the *Item Description* window

QUESTION 2.2

What is the spatial resolution, or compilation scale, of *SC_Dtl_Counties*?

a. 1:1,000
b. 1:10,000
c. 1:100,000
d. 1:1,000,000

QUESTION 2.3

What is the spatial resolution, or compilation scale, of *SC_Gen_Counties*?

a. 1:2,000
b. 1:200,000
c. 1:2,000,000
d. 1:20,000,000

QUESTION 2.4

Both *SC_Gen_Counties* and *SC_Dtl_Counties* cover the same area, but they are compiled at very different scales.

a. Which of these two layers is more generalized? _____

b. Which of these two layers is the larger-scaled layer? _____

QUESTION 2.5

When compared to the spatial resolution of the other data layers, the county layer that will be the most appropriate choice to use given its resolution will be *SC_Gen_Counties*.

a. True
b. False

Setting the Map Scale

CONCEPT

Map scale in **ArcMap** is a very fluid parameter; you can zoom in and zoom out both on the data and on the page on which the data resides. When you are adding symbols and labels, however, or preparing a layout for printing, you will often want to have more control over how map scale behaves.

You can, for example, set a *reference scale* for your data frame, which will freeze symbol and text sizes so that they always match the scale, regardless of whether you zoom in or out. You might also want to set a *fixed scale* for your data frame; this setting keeps the map scale of your layout constant – you aren't allowed to zoom in or out on the data itself, but you can zoom in or out on the page.

Reference Scale

1. Click the *Full Extent* tool on the *Tools* toolbar to reset the view of *South Carolina*
2. Right-click the county layer you will not use in making your map and choose *Remove*
3. Right-click the remaining county layer and choose *Properties*, then the *Symbology* tab
4. Click on the *Symbol* to change the county colors—choose a *fill color* and an *outline color* of your choice and click *OK* then click *OK* again
5. Right-click the county layer again and choose *Label Features*
6. Using the **Data View** Zoom tools, zoom in and out several times on the counties, noting how the labels change size
7. Right-click the data frame and choose *Properties*, then the *General* tab
8. Change the name of the data frame (your map) to **SC Reference**
9. Set the *Reference Scale* for the map to **1:2,000,000** and click *OK*

QUESTION 2.6

Zoom in and out on the map again, paying attention to how the labels behave.

By setting the reference scale, the county labels automatically size up or down relative to their surrounding features as you zoom in or out.

a. True
b. False

Fixed Scale

10. Click on the *Layout View* icon in the bottom left-hand corner of the Data View

The large rectangle with the shadow behind it is your *page* (Figure 2.4). The dashed line just inside that rectangle shows you how much of that page will actually print. The rectangle with the light blue rectangular handles holds your map in it and is your *data frame* or map container, which is currently named *SC Reference* in your TOC.

11. On the Main Menu bar, click *File > Page and Print Setup*
12. Change the *Orientation* to **Landscape**, then click *OK*
13. Click on the data frame to select it, then right-click and choose *Properties*
14. Select the *Size and Position* tab and set the *Width* to 9 and the *Height* to 7; click *OK*
15. Select the **SC Reference** data frame and use the drop-down list of scales in the Standard Toolbar to choose a scale in which the state of S.C. will start to fill the data frame
16. Use the **Data View** zoom out tool and zoom out
17. Reset the scale to its previous value
18. Right-click the data frame, then click *Properties*

FIGURE 2.4

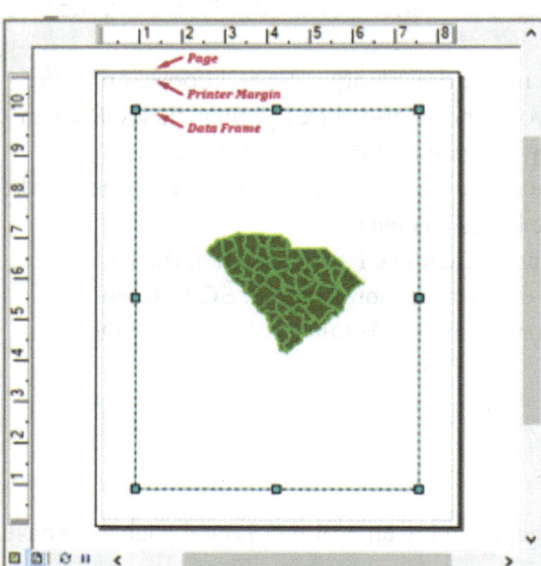

19. Select the data frame tab, then in the *Extent* window, click the drop down box and select *Fixed Scale*
20. Set the scale to **1:2,500,000**, then click *OK*

Go back to the **Data Zoom** tools. Notice that once you have set the fixed scale for your map document, you can no longer zoom in and out on your data.

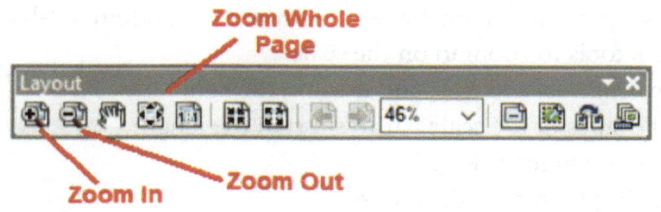

There are a second set of zoom tools, however, built specifically for the *Layout View*. If you don't see these tools, go to the *Customize* menu and select *Toolbars – Layout* to make them visible.

21.   Using the **Layout** Zoom tools, zoom in and out several times

QUESTION 2.7

What happens when you use the *Layout Zoom* tools as opposed to the *Data View* zoom tools?

a.   The Layout Zoom tools change the scale of the data layers
b.   The Layout Zoom tools perform the same function as the Data View Zoom tools
c.   The Layout Zoom tools zoom in and out on the map layout instead of the data layers
d.   The Layout Zoom tools zoom in and out only on the map legend and map title

22.   Use the **Layout View** *Zoom Whole Page* tool to make the entire page fill the display
23.   From the *File* menu, choose *Save* to save your updated project

Creating Scale Representations

CONCEPT

Mathematically, a map's scale is a *ratio* and is defined as *Map Distance/Ground Distance*. This ratio, when Map Distance is represented by a measurement of 1 and both map and ground distances use the same units of measurement, is called a *representative fraction* (RF). Using conversions, we could change the units of measurement for the *Ground Distance*, and create a *verbal statement* of scale. 1:24,000, for example, could be restated as "1 inch represents 2,000 feet." Another user-friendly format for scale is the *graphic scale*. A graphic scale is simply a line on the map that is subdivided into major units representing ground distances.

1. Click *Insert* on the Main Menu bar, then click *Scale Bar*
2. In the Scale Bar Selector window, click on *Alternating Scale Bar 2*, then click *OK* (Figure 2.6)
3. Click and drag the scale bar into the white space toward the bottom left-hand corner
4. Use the *Layout Zoom* tools to zoom in on the scale
5. Select the *Select Elements tool* and click and drag on the light blue anchor points on the sides of the scale bar to shorten the length, making the scale bar 100 miles in length
6. Right-click the scale bar, then click *Properties*
7. Select the *Scale and Units* tab if it is not already selected
8. Change the *Number of divisions:* from the default of 4 to 2
9. Change the *Number of subdivisions:* from the default of 4 to 2, then click *Apply*
10. Click *OK*

Select Elements Tool

FIGURE 2.6

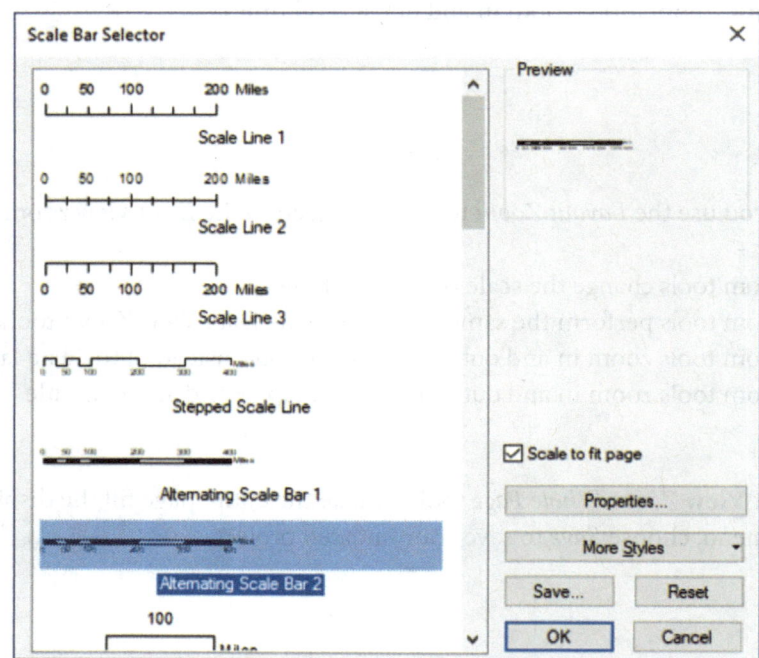

QUESTION 2.8

Convert the RF of 1:2,500,000 to a verbal statement where "1 inch represents approximately x miles." Which of the following statements would be the correct result?

a. 1 inch represents approximately 3.9 miles
b. 1 inch represents approximately 39 miles
c. 1 inch represents approximately 325 miles
d. 1 inch represents approximately 3500 miles

Creating Effective Symbols

Verifying an Attribute's Level of Measurement

CONCEPT

Usually, you are interested in a specific attribute attached to the real-world feature you are symbolizing. The way in which that attribute has been measured directly affects how we may visually present its variation on the map. In the context of levels of measurement, four measures affect how we may manipulate a symbol: *nominal*, *ordinal*, *interval*, or *ratio*.

1. Right-click *SC_MjrCities*, then click *Open Attribute Table*

Three of the attributes in this table include RANK_MAX, Capital, and GN_POP. RANK_MAX is used to record the ranking of each city within the state according to its metropolitan population; Capital is used to record whether or not the city is a state capital; and GN_POP is used to record the urban population of each city.

QUESTION 2.9

For each of these attributes, identify their level of measurement:

RANK_MAX   _____

Capital   _____

GN_POP   _____

2. Close the attribute table

Selecting a Visual Variable for Your Symbol

CONCEPT

A symbol's dimension and its attribute's level of measurement affect which *visual variable* should be used in drawing the symbol. Bertin's six traditional visual variables are: shape, size, hue, value, orientation, and texture.

QUESTION 2.10

You have been asked to make three maps, each showing the cities of South Carolina. One map should show how the cities vary by metropolitan population rank (RANK_MAX), one should show how they vary by urban population (GN_POP), and the third should show whether or not they are a capital city (Capital). For each of these cases, list the most appropriate visual variable to manipulate to showcase this variation when using a point symbol to represent the cities.

RANK_MAX   _____

GN_POP   _____

Capital   _____

3. Right-click the layer that displays your counties in the *TOC* and turn off the county labels
4. Make the *SC_MjrCities* layer visible
5. Right-click the *SC_MjrCities* layer, then click *Properties* and the *Symbology* tab (Figure 2.7)
6. Click to select *Quantities*, then *Graduated Symbols*
7. For the *Fields Value:* click the drop-down box and choose GN_POP
8. Change the *Symbol Size from:* **12** to **36**
9. Select *Template* if you want to change the color of the symbols
10. Click *Label*, then *Format Labels*
11. Under *Rounding*, change the number of decimal places to **0**
12. Click *OK*
13. Click *OK* again
14. Right-click *SC_MrjCities* and choose *Properties*, then the *Labels* tab

FIGURE 2.7

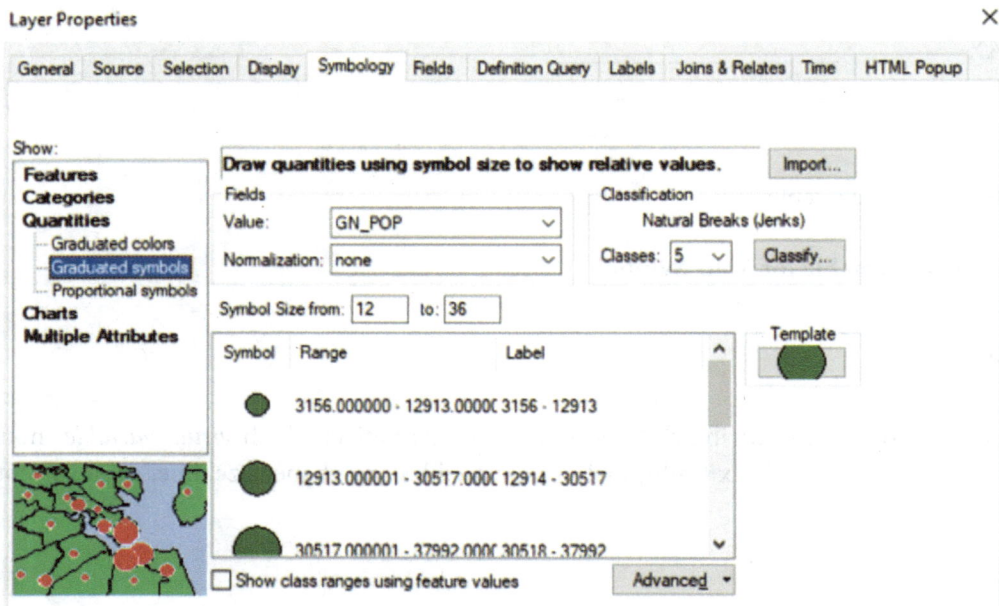

15. Change the label size to **14** and the label style to Bold (**B**)
16. Click *OK*
17. Right-click *SC_MrjCities* and choose *Label Features*

QUESTION 2.11

What visual variable do graduated symbols use to display variation in city rankings?

a. Size
b. Value
c. Texture
d. Hue

Setting a Data Classification

CONCEPT

Classification refers to grouping similar attribute values so that you can reduce the complexity of the final map. Instead of trying to display each value in a dataset with a unique symbol, the cartographer uses a smaller set of symbols to display a limited number of groups that contain similar values.

1. Make the *SC_MjrRds* layer visible

The **RTTYP** field in this layer's attribute table is a ranking of roads according to their importance. This field can be used for classification, and will allow you to more easily distinguish between each of these route types visually.

2. Right-click the *SC_MjrRds* layer, click *Properties*, then the *Symbology* tab
3. Click *Categories*, then select *Unique values*
4. Change the *Value Field* to **RTTYP**, then click *Add All Values* (Figure 2.8)
5. Double-click each symbol line width and change its symbology to create a hierarchy of roads for the map. Use the following symbology, changing the colors as you see fit:
    For I, choose **Freeway**, with a width of 3.4
    For U, choose **Major** Road, with a width of 1.5
    For M, choose **Major** Road, with a width of 1.5
    For S, choose **Major** Road, with a width of 1.5
6. Click *OK*
7. Use the Layout *Pan* tool to re-center your map within the Layout View
8. Use the Layout *Zoom* tools to make sure your entire map shows in the Layout View
9. From the *File* menu, choose *File-Save*
10. From the *File* menu, choose *Export Map*

FIGURE 2.8

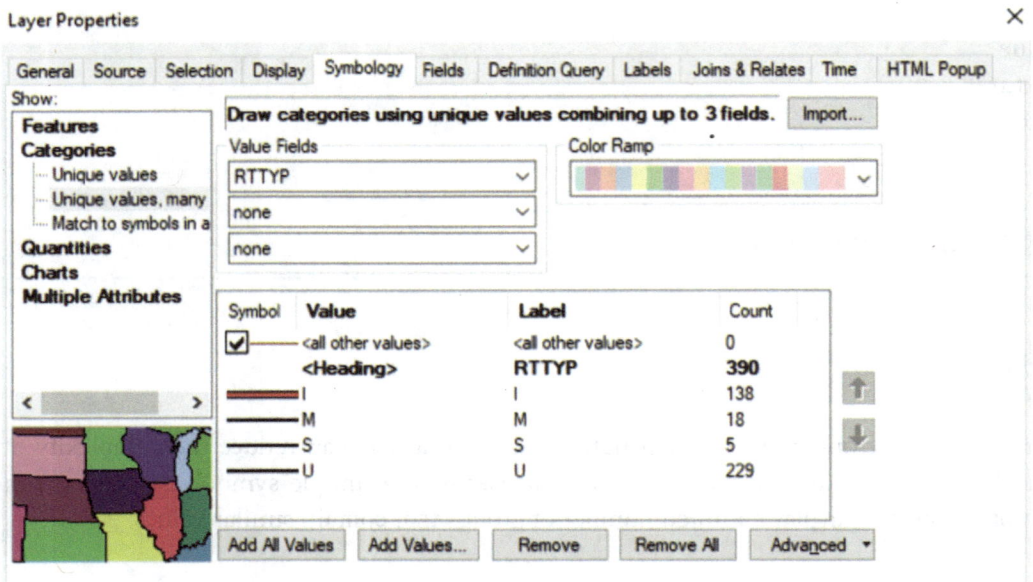

11. In the *Save In* drop-down box at the top of the dialog, navigate to the same location that you saved the original map document

12. Under the *Save as Type* drop-down box, choose *PDF*

13. Name your file to match the name of the original map document minus its extension, then click *Save*

14. Double-click the *Computer* icon your desktop

15. Navigate to the location where you think you saved your file and verify that it is there

Note: To receive full credit during the grading process, the map you export should reflect the latest changes as directed by the exercise.

16. Exit **ArcMap**

17. If you are on a machine in a lab and are finished with your session, log off and/or restart your machine before you leave.

Chapter 3

The Geographic Coordinate System

The foundation upon which GIScience and its related technologies is built is ancient; many of the initial ideas were conceived by Greek scholars. We know this because *Strabo*, a Greek geographer born around 63 B.C., reported on these ideas and the people behind them. Strabo spent five years at the Library of Alexandria in Egypt studying these writings. His 17-volume *Geography* combined knowledge from mathematics, physics, politics, and history into a discourse on geography; it is one of the few writings to survive the Greek period (Brown, 1977). In his research for *Geography*, Strabo discovered that Greek philosophers began to influence geographic thought as early as 600 B.C. Of their many contributions, there are three key concepts we need to explore in this chapter: 1) the concept of the Earth as a sphere; 2) the determination of the circumference of the Earth; and 3) the origination of the geographic coordinate system, or latitude and longitude. Each of these ideas undergirds all of GIScience.

The Earth's Shape

We can find evidence through Strabo's writings that *Pythagoras* was one, if not the first, to contemplate both the shape and size of our planet. Certainly one of the more famous Greek philosophers, Pythagoras lived during the 6th century B.C. He founded a school of philosophy around 523 B.C., and proved several scientific hypotheses, including the Pythagorean Theorem. Of the hypotheses that Pythagoras put forth, potentially the most important one was his proposal that *the Earth is spherical in shape.*

Although it remains unclear exactly how he came to support this contention, *Aristotle* (384-322 B.C.), another of the founding figures in Western philosophy, later agreed with him. More importantly, Aristotle used *celestial observations* to prove the idea that our planet was, indeed, spherical. For example, during a partial eclipse of the moon he noted the curve of the Earth's outline on the moon. Since the alignment of the moon, Earth, and the Sun caused the eclipse, he argued that our planet must be spherical. He also observed that certain stars visible in Egypt and Cyprus were not visible at locations further north, and again attributed this to a curving, not straight, horizon (Brown, 1977).

Related to these ideas, and a simple method for observing the spherocity of Earth, is to observe what happens when a ship moves out to sea from port, as described by Campbell (1998). As the ship increases its distance from us, it doesn't just appear then disappear, as it might if our planet were flat or disc-shaped. Rather, it first grows smaller, and then appears to sink, with the hull disappearing first and the top of the mast disappearing last. This phenomenon occurs uniformly over the planet, providing further evidence for a spherical shape. Finally, we in the 21st century have what is undoubtedly the most direct evidence – photographs of our planet by astronauts and satellites. One of the most famous is the "Blue Marble", taken by astronauts aboard *Apollo 17* in 1972 (Figure 3.1). This image shows a fully illuminated Earth; its name reflects the perception of the astronauts that the Earth looked like a glass marble.

FIGURE 3.1
NASA's 1972 Blue Marble image.
Image: *NASA's Earth Observatory.*

The Earth's Circumference

With Aristotle's observations to back up the theory of a spherical Earth, the general population soon came to accept the idea, and people began to wonder just how large a sphere Earth might be. Several made guesses, including Aristotle, but no one supplied any details on how they arrived at their estimates. The first person to lay out these ideas using a logical framework was the same man that coined *geography*: *Eratosthenes*. Eratosthenes' method and resulting accuracy in estimating the circumference of our planet is one of the great accomplishments of Greek science and a quite interesting story (Thrower, 1996).

Eratosthenes began by accepting the Earth as a sphere. He knew that if he were to "slice" this sphere in half, he would have a circle (Figure 3.2a). Using some basic geometry and a few facts about nearby locations, he proceeded to set up a method for estimating the circumference. First, he divided the circle, much the way you might section a grapefruit. At this point, he grappled with more questions: how many sections comprised the circle, and what was the length of arc for each section?

FIGURE 3.2

Eratosthene's method for establishing the circumference of Earth. Source: Elisabeth Nelson

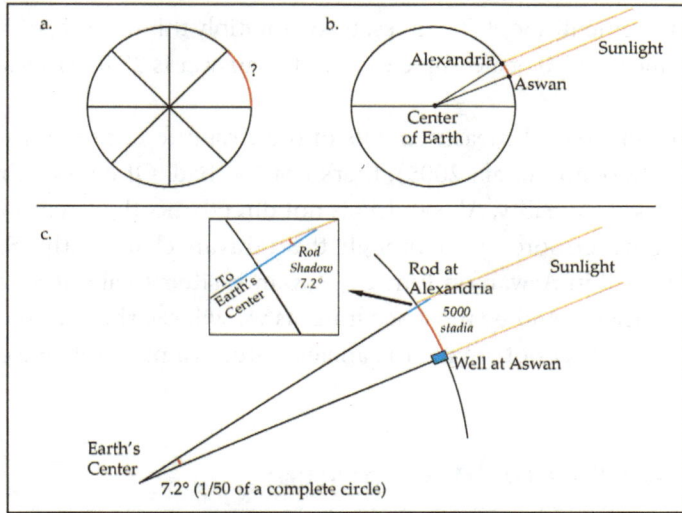

To answer these questions, Eratosthenes knew he needed real-world data. As luck would have it, at least according to one story, a caravan passed through Alexandria with a report about a well they saw in Aswan, Egypt (Laskey, 2001). According to the travelers, on June 21, at noon, the Sun shone directly into the well – casting absolutely no shadows. Eratosthenes thought about this, and realized that on that same day, there *were* shadows cast at noon in Alexandria, Egypt. This seemed curious to him, and the more he thought about it, he realized—with a little help from the Sun and a spherical Earth—he could set up a geometry problem to solve his puzzle.

He began by making two assumptions:
1) The rays coming from the Sun were parallel as they struck the Earth
2) Alexandria was directly north of Aswan – in other words they were both located on the same circle that would be created if the Earth were sliced in half along that route (Figure 3.2b).

The laws of geometry then say the angle at which the rays of light hit a straight line, such as a rod sticking straight up out of the ground, equals the alternating angle at the Earth's center whose two rays pass through the endpoints of the arc between Alexandria and Aswan. When Eratosthenes measured the angle of the shadow cast by the Sun in Alexandria, he discovered it was approximately 7.2 degrees (Figure 3.2c). At this point, he had one part of his puzzle solved. It was about 7.2 degrees from Alexandria to Aswan.

Now, he needed to convert that into a practical measurement, requiring him to make some additional assumptions. First, he needed to know what percentage of the circle 7.2 degrees equaled. This was easy; he knew a circle was comprised of 360 degrees, so if you divided that by 7.2, you would come up with about 1/50 of the complete circumference. His last missing piece was to come up with the actual distance from Alexandria to Aswan. He arranged to have surveyors trained to walk with equal steps and then actually walk the distance between these two cities; they estimated the distance to be 5000 stadia (a stade was equivalent to the length of a Greek stadium). Today, scientists consider a stade about 1/10 of a mile, making the distance that they walked, in our terms of measurement, about 500 miles. If you multiply this distance of 500 miles by 50, since the distance is 1/50 of a complete circle, the answer is 25,000 miles.

Considering our most recent measurements of the circumference of our planet puts it at 24,906 miles (Slocum, et al., 2005), that's not too bad. Of course, Eratosthenes was also a little lucky! In reality, Alexandria is not directly north of Aswan, but about 3 degrees west. Furthermore, even though the caravan claimed the Sun cast no shadow down the well in Aswan on June 21, Aswan is situated about 37 miles north of the point where that would be most accurate, as we will see shortly. In the end, the mistakes cancel each other out, giving the ancient Greeks a most reliable estimate.

The Geographic Coordinate System

Eratosthenes wasn't just interested in determining the circumference of the Earth as an academic exercise. He was actually working on a map of the habitable world and wanted to know the length of a degree to plot locations from the sphere to a flat piece of parchment with some degree of accuracy. To do that, he needed a systematic way of dividing the Earth that would also allow him to plot locations of places in two dimensions.

He began by following some of the work of earlier Greek scholars, and divided his map into ". . . unequal, straight-sided geometrical figures compatible with the shapes of different countries" (Thrower, 1996:22). This was largely unsuccessful, but it did prompt *Hipparchus* (190-120 B.C.), one of the best astronomers of the time, to propose a systematic grid of intersecting lines that eventually became our *geographic coordinate system* of today. As an astronomer, Hipparchus' idea was to project the grid lines on the sphere using the rotation of the Earth and the movement of the Sun as a more scientific and rigorous underpinning.

He started with a core concept in astronomical studies: the *plane of the ecliptic* (Figure 3.3). The plane of the ecliptic is the plane in which the Earth orbits the Sun. As Earth completes its annual orbit around the Sun, it rotates on its polar axis. If this axis was perpendicular to the plane of the ecliptic, the Sun's rays would always shine directly on the *equator*, which is an imaginary line established when you pass a plane through the surface of Earth perpendicular to its axis of rotation (Figure 3.4). Imaginary lines such as the equator, which bisect the planet into two equal halves, are known as *great circles.* The equator bisects our planet into Northern and Southern Hemispheres. The equator provides us with the starting point for determining latitude, or position north or south on the globe. Other lines of latitude may also be plotted; they are known as *small circles,* since they do not bisect Earth. The key to locating these lines lies in the fact that the Earth's axis of rotation is *not* perpendicular to the plane of the ecliptic, but, in fact, is tilted approximately 23.5° relative to the perpendicular.

FIGURE 3.3
Plane of the ecliptic and the four seasons.

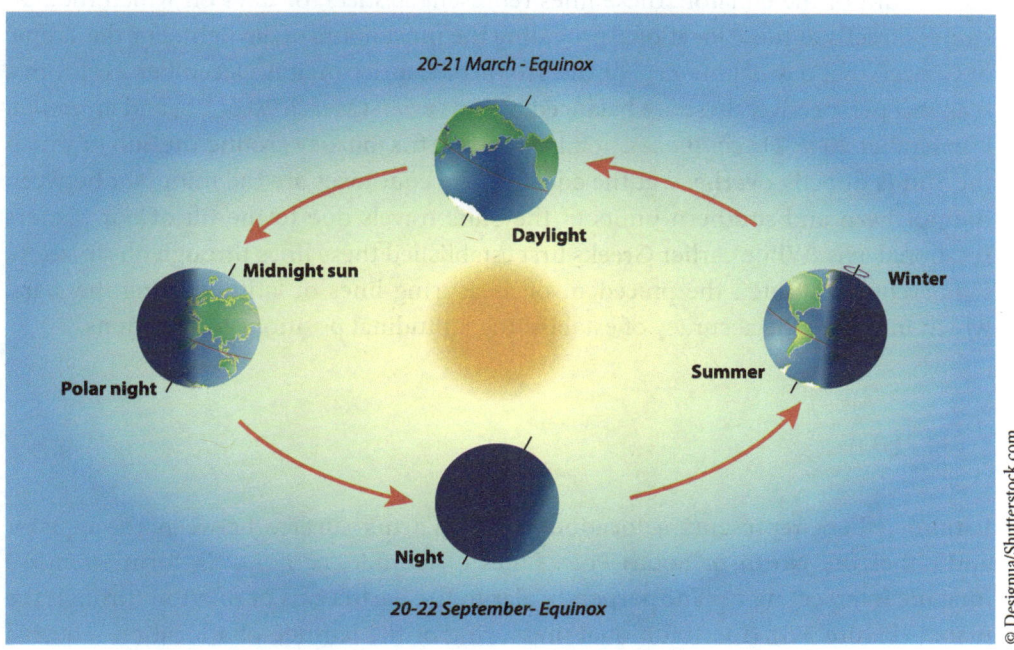

FIGURE 3.4
Framework for the system of latitude.

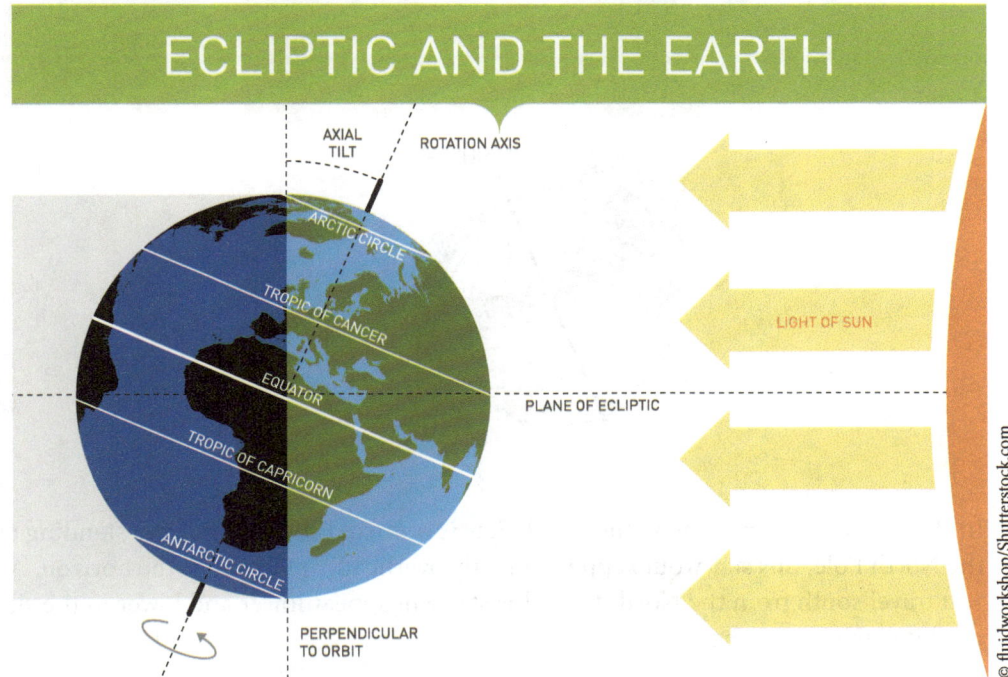

As Figure 3.3 shows, with the Earth's axis tilted relative to the plane of the ecliptic as it makes its orbit around the Sun, the Sun's rays do not always shine perpendicularly on the equator, but travel north and south throughout our year. Changes in the Sun's position relative to the equator can be measured using the slant and length of shadows. This type of tracking by the Greeks led to the development of two further lines of latitude relative to the equator and shown in Figure 3.4: the *Tropic of Cancer* (Summer Tropic), at 23.5° north of the equator, and the *Tropic of Capricorn* (Winter Tropic), at 23.5° south of the equator. These lines represent *solstices*, or days on which the Sun shines directly at these locations, providing the most hours of daylight. For the Tropic of Cancer, that day is June 21. For the Tropic of Capricorn, it is December 21. From a seasonal perspective, these are balanced by the *vernal* (March 20 or 21) and *autumnal* (September 20 or 21) equinoxes, points in the Earth's journey around the Sun in which the Sun is directly overhead at the equator. The equinoxes are the midpoint between the northern and southern limits of the Sun's travels due to the tilt of our planet's rotational axis. While earlier Greeks first established these lines through observation, Hipparchus instituted the precedent of measuring lines of latitude using the stars, which increases the accuracy of establishing latitudinal positions for locations.

LATITUDE

Latitude, then, represents a location on the Earth's surface between the equator and either the North or South Pole. Lines of latitude, or *parallels*, form when we imagine intersecting a plane perpendicular to the Earth's axis of rotation through the planet (Figure 3.5). The traditional measuring of the latitude of a location required determining the ". . . function of the angle between the horizon and the North Star (or some other fixed star)" (Dent, et al., 2009:28).

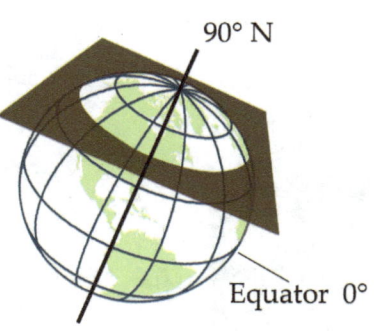

FIGURE 3.5
Visualizing lines of latitude.
Data: *Made with Natural Earth. Free vector and raster map data @ naturalearthdata.com.*
Source: Elisabeth Nelson

90° N

Equator 0°

In the Northern Hemisphere, the North Star is our fixed star. If you were standing at the North Pole, this star would appear directly overhead, or 90° above the horizon. As you travel south from the North Pole, this star will appear lower and lower in the sky until it reaches 0° at the equator.

To determine the angle of latitude for any given position in the Northern Hemisphere, we can use the position of the North Star in the sky at our location and a little geometry (Muerchke, et al., 2001:288). The proof requires that you assume the line of sight from the center of the Earth to the North Star is parallel to the line of sight from your position on the planet to the North Star (Figure 3.6). Once this is accepted, the laws of geometry say that a straight line drawn from your position on Earth through the Earth's center will intersect those parallel lines of sight at the same angles (55°). Because of this, the adjacent angles (35°) must also be equal, since the sum of adjacent angles in both cases must equal 90°. 35° represents your latitude, which can also be measured via surveying or GPS technology today.

FIGURE 3.6

Measuring latitude.
Data: *Made with Natural Earth. Free vector and raster map data @ naturalearthdata.com.*
Source: Elisabeth Nelson

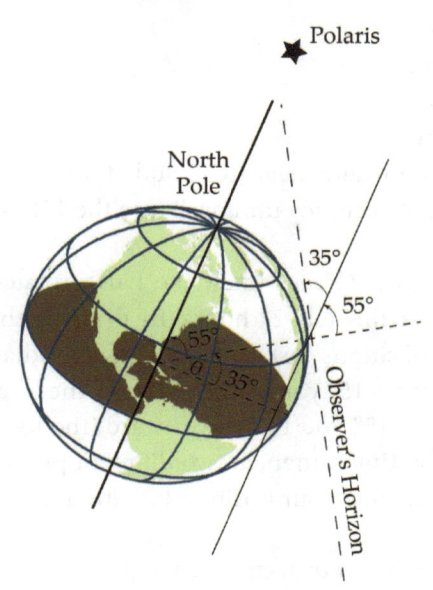

The range of latitude varies from 0° at the Equator to 90° at each pole, with the parallels marking the change of degree as you approach either pole from the equator. For example, Cairo, Egypt and New Orleans, Louisiana are both located at approximately 30° N. If we wanted, however, we could be even more accurate by measuring parts of a degree, i.e., using minutes (60 in a degree) and seconds (60 in a minute). In this case, Cairo's latitudinal coordinate would be 30° 03′ 53″ N and New Orleans would be 29° 57′ 15″ N. This unit of measurement is derived from the *sexagesimal system* of dividing a circle into 360 degrees, and is frequently denoted as a *DMS format* (degrees, minutes, seconds) in GIS.

LONGITUDE

Of course, just knowing the latitude of a location is only part of the equation for specifying position on our planet. If we know that both Cairo and New Orleans are at 30° N, but don't know where they are positioned on the globe east or west, we have no idea where along that line of latitude the cities actually exist. To pinpoint the exact location on the globe, we also need to know each location's *longitude*. We create lines of longitude by passing a plane through the center of the Earth so that the plane intersects the Earth's axis of rotation and bisects the planet into Eastern and Western Hemispheres. Because they bisect the Earth, they, like the equator, also form great circles (Figure 3.7). You should note that lines of longitude are fundamentally different from lines of latitude. Lines of latitude are parallel to one another and spaced at equal intervals on the globe. Lines of longitude, on the other hand, are equally spaced around the earth at the equator, but the distance between them decreases as you move north or south because they intersect the axis of rotation, converging at the poles.

This difference creates a dilemma – if all lines of longitude are great circles, *which circle do you use as a starting point for measuring location east and west?* The answer is not an easy one. Early on, each country tended to have their own *Prime Meridian* – the line of longitude from which they measured distance in degrees from east and west – and typically, it ran through their capital city. That was fine when all anyone wanted to map was their own country, but as you might imagine, it created quite the mess when countries wanted to communicate locations with each other. The other issue with longitude was how, precisely, to measure it from any given starting point. From a conceptual standpoint, the Greeks were able to reason it out quickly, but the actual ability to technically measure it eluded them; the problem wasn't actually solved with any degree of accuracy until well into the 1700s.

FIGURE 3.7
Visualizing lines of longitude.
Data: *Made with Natural Earth. Free vector and raster map data @ naturalearthdata.com.*
Source: Elisabeth Nelson

At the root, longitude is really a time problem; the Greeks knew the Earth rotated west to east, and that a full revolution took approximately 24 hours. By dividing the 360° of the sphere by 24 hours, they knew that positions east or west of any meridian changed by 15° every hour. In other words, for every 15° you travel eastward, the local time moves ahead one hour. Likewise, for every 15° you travel westward, the local time moves back one hour. If you know the local times, then, at two different points, you can use the difference between them to determine your position longitudinally.

Unfortunately, the Greeks didn't have a mechanism sturdy or accurate enough to measure time from one location to the next. Over the centuries, many people worked at developing a timepiece. It wasn't until 1773 that a *marine chronometer*, a timepiece developed by *John Harrison*, was accurate enough to be acceptable for navigation (Brown, 1977). After that, it was another century or so, in 1884, before the international community would formally designate a world Prime Meridian. In that year, the International Meridian Conference chose the meridian (1/2 a line of longitude) that ran through *Greenwich, England* to be the world's standard from which to measure longitude.

By setting the chronometer to the time in Greenwich, England, and then comparing the time on the chronometer to local high noon, mariners could use the difference between the times at the two locations to determine the ship's longitude. Longitude values, thus, run from 0° to 180° east and 0° to 180° west from the Prime Meridian, with the other half of the line of longitude being designated as the *International Date Line* (Figure 3.8). This line separates consecutive calendar days, with the Eastern Hemisphere always being one day ahead of the Western Hemisphere. With the other half of the coordinate system in place, we can now fully distinguish between the locations of Cairo (30° N 31° E) and New Orleans (30° N 90° W). This spherical coordinate system of latitude and longitude is known as the geographic coordinate system.

FIGURE 3.8
Measuring longitude.
Data: *Made with Natural Earth. Free vector and raster map data @ naturalearthdata.com.*
Source: Elisabeth Nelson

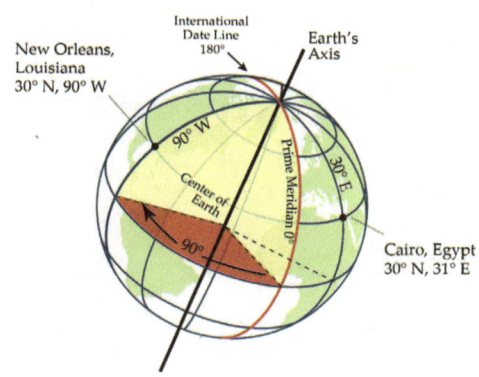

References

Brown, L.A. 1977. *The Story of Maps.* Dover Publications, Inc.: New York, New York

Campbell, J. 1984. *Introductory Cartography.* Prentice-Hall, Inc.: Englewood Cliffs, New Jersey.

Campbell, J. 1998. *Map Use & Analysis.* 3rd ed. WCB McGraw-Hill: Boston, Massachusetts.

Crane, G.R. 2009. *Perseus Digital Library Project.* Tufts University http://www.perseus.tufts.edu. Last accessed January 5, 2009.

Dent, B.D., Torguson, J.S., and T.W. Hodler. 2009. *Cartography: Thematic Map Design.* 6th ed. McGraw-Hill: New York, New York.

Laskey, K. 2001. *The Librarian Who Measured the Earth.* Houghton Mifflin: Morris Plains, New Jersey.

Muehrcke, P.C., Muehrcke, J.O., and A.J. Kimerling. 2001. *Map Use: Reading, Analysis, Interpretation.* 4th ed. JP Publications: Madison, Wisconsin.

NASA-Johnson Space Center. 2009. *The Gateway to Astronaut Photography of Earth.* http://eol.jsc.nasa.gov/scripts/sseop/photo.pl?mission=AS17&roll=148&frame=22727 Last accessed January 5, 2009.

Raisz, E. 1948. *General Cartography.* McGraw-Hill Book Company, Inc.: New York, New York.

Robinson, A.H., Morrison, J.L., Muehrcke, P.C., Kimerling, A.J., and S.C. Guptill. 1995. *Elements of Cartography.* 6th ed. John Wiley & Sons, Inc.: New York, New York.

Slocum, T.A., McMaster, R.B., Kessler, F.C., and H.H. Howard. 2005. *Thematic Cartography and Geographic Visualization.* 2nd ed. Pearson Education, Inc.: Upper Saddle River, New Jersey.

Thrower, N.J.N. 1996. *Maps and Civilization: Cartography in Culture and Society.* University of Chicago Press: Chicago, Illinois.

Exercise 3
Latitude and Longitude in ArcGIS

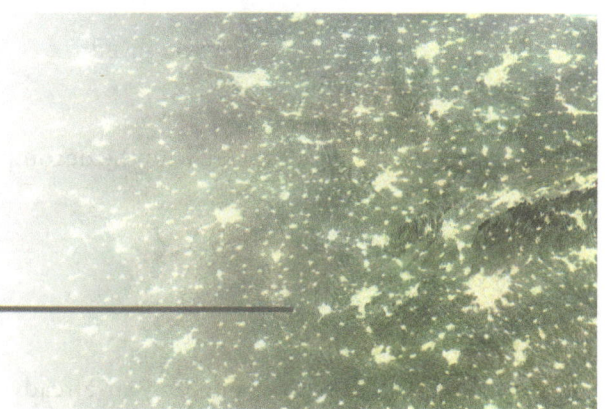

In Exercise 3, you will explore how latitude and longitude coordinates are used within a GIS.

OBJECTIVES

- Explore the system of latitude and longitude
- Estimate how latitudinal and longitudinal distances vary over our planet
- Create a layer from a table of latitude and longitude coordinates

SOFTWARE INFORMATION

Introducing ArcMap uses ESRI's **ArcGIS 10.6** software. **ArcGIS** is a commercial GIS package, available in geospatial labs affiliated with schools that have a campus-wide site license for the software. Instructors at these campuses may also request 1-year student versions of the software at http://www.esri.com/landing-pages/education-promo.

DATA

The data for the exercises is available from the Kendall Hunt Student Ancillary site. See the inside front cover for access information. You may also be directed to download the data from a different location by your instructor.

1. Download the data as instructed by your instructor
2. Save the file to a location where you have read/write privileges (USB key, home computer, class server)

The file you just saved is in a compressed (*.zip) format, and was created using **7-zip** freeware (http://www.7- zip.org). To use the data for this exercise, you must decompress it using the same software.

3. Locate the zipped file that you saved
4. Right-click the file
5. Choose *7-Zip - Extract Files*
6. Click *OK* to create a folder with the decompressed data

Specify a Home Folder

In this exercise, rather than opening an already existing ArcMap map document, you will be creating one from scratch. Because of this, you will also need to explicitly associate a Home Folder and a default geodatabase for your new map. This requires a few steps up front, but will make things easier later.

1. Start **ArcMap** by double-clicking on your desktop icon or by clicking >All Programs > ArcGIS > **ArcMap 10.6**
2. In the *ArcMap - Getting Started* dialog box, click *Cancel*
3. If the *Catalog* window is not open, click on the *Catalog* tab on the right side of the ArcMap interface OR go to *Windows – Catalog* to open it
4. Click the pushpin in the upper right corner of the *Catalog* window to freeze the window in place (Figure 3.1)

FIGURE 3.1

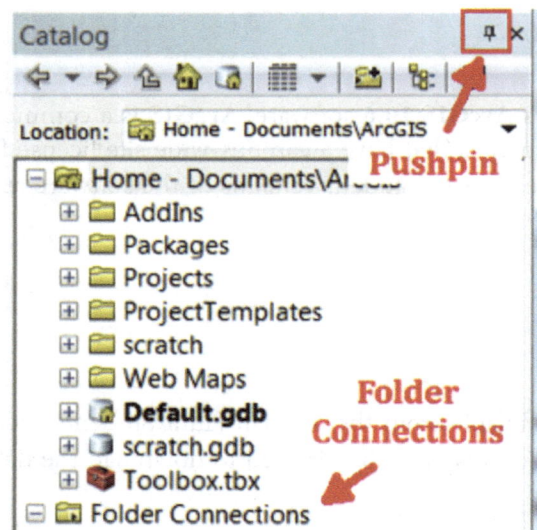

Notice that the default Home folder is set *Home-Documents\ArcGIS*. You want your Home folder to be *Chapter 03*, which is where your data for this exercise resides.

5. From File, choose *Save As*
6. Name your file using this convention: *lastname_Exercise03*
7. Save your map to the *Chapter03* folder containing *Exercise_03.gdb*

Your home folder should now be updated to the correct location. If you have any questions, verify this with your instructor before proceeding.

Specify a Default Geodatabase

It is also important to link your map document specifically to the geodatabase that houses your data for the exercise. With this link in place, any changes or additions you make to your data will be stored in the correct location on the computer.

1. From the *File* menu, choose *Map Document Properties*
2. Click to select the file folder button to the right of the default geodatabase path
3. Browse to find *Exercise_03.gdb*
4. Click to select it, then click *Add*
5. Next to pathnames, place a checkmark by *Store relative pathnames to data sources*
6. Click *OK*

Add Spatial Data to the Map Document

To view the spatial data in your geodatabase, you must create links to it from within the map document.

1. In the Catalog, find your *Home* folder
2. Click the + sign to the left of *Exercise_03.gdb* to see the spatial data files you will use in this Exercise
3. Click and drag each data layer in the geodatabase to your *Data View* window

REFINE THE DEFAULT DATA VIEW

4. In the *TOC,* click on the boxes next to *Graticule_1_Degree_Increments, World_Cities,* and *Tectonic_Plates* to make the layers invisible
5. In the *TOC,* double-click on *Layers* to bring up the data frame properties dialog box
6. Under the *General* tab, change the data frame name to **World**, then click *OK*
7. In the TOC, right-click on *Graticule_20_Degree_Increments* data layer and select *Properties*
8. Click on the *Labels* tab to set up labels for a graticule on your map
9. Set the *Text String Label Field* to **degrees**, then click *OK* (Figure 3.2)
10. Right-click the *Graticule_20_Degree_Increments* layer again, this time choosing *Label Features*

FIGURE 3.2

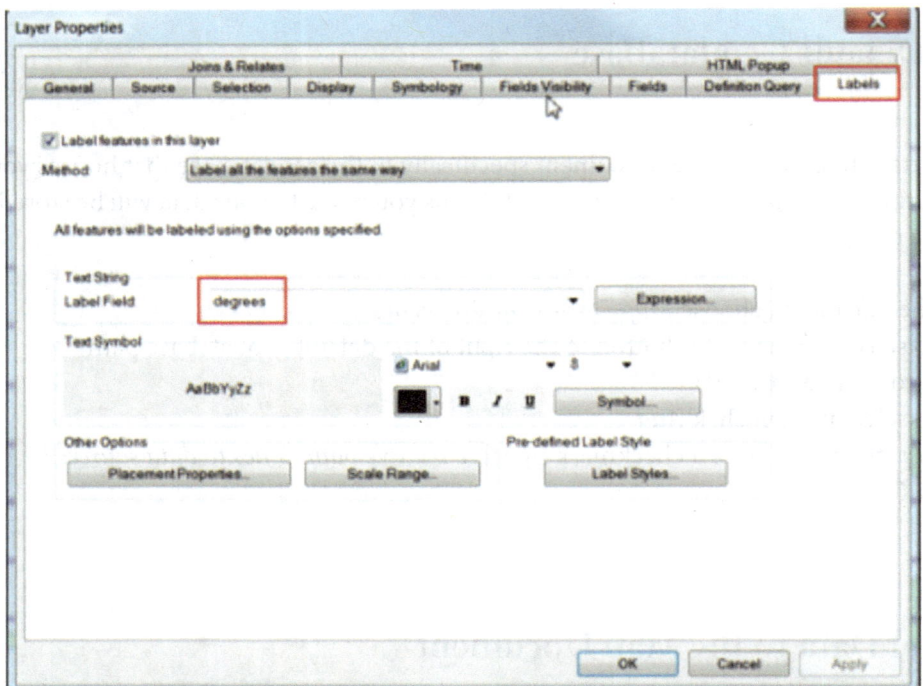

Latitude and Longitude

We specify feature locations on Earth by latitude and longitude, which forms a network of intersecting lines called a *graticule*. Latitude is expressed as 0° at the Equator and runs to +90° at the North Pole and -90° at the South Pole. Longitude is 0° at the Prime Meridian, and runs to +180° to the east and -180° to the west, meeting at the International Date Line.

1. In Data View, move your cursor over the map and watch the latitude and longitude values track in the lower right corner of the ArcMap window

Locating Features

QUESTION 3.1

Use the *Zoom* and *Pan* tools to find the locations listed below. When you have found the location, use the *Identify* tool to determine the country located at that intersection (be sure not to click directly on a graticule line):

a. 30° North, 30° East: _____

b. 20° South, 60° West: _____

c. 75° North, 50° West: _____

d. 20° South, 45° East: _____

QUESTION 3.2

1. Click on the check box next to *World_Cities* to view the data layer

Use the *Find* tool to locate the following cities, then, in the results window at the bottom of the *Find* tool dialog box, right-click the city and use *Identify* to determine its latitude and longitude:

a. Amsterdam: _____

b. Canberra: _____

c. Casablanca: _____

d. Sucre: _____

e. Monterrey: _____

CONVERTING FROM DMS to DECIMAL DEGREES

QUESTION 3.3

The following cities are located using the traditional latitude/longitude degree system:

a. Athens, Greece: 37° N, 23° E

b. Auckland, New Zealand: 36° S, 174° E

c. Buenos Aires, Argentina 34° S, 58° W

d. Caracas, Venezuela 10° N, 66° W

Specify their locations in *decimal degrees* (remember that longitude is listed first here):

    a.   Athens, Greece: _____

    b.   Auckland, New Zealand: _____

    c.   Buenos Aires, Argentina _____

    d.   Caracas, Venezuela _____

1. Click on the check box next to *World_Cities* to make the data layer *invisible*
2. Right-click on *Graticule_20_Degree_Increments* and choose *Properties*
3. Under the *Symbology* tab, click on the symbol and change the line *Width* to **2**
4. Click *OK*, then click *OK* again

The Geographic Coordinate System (GCS)

In a GIS, latitude and longitude coordinates are part of the geographic coordinate system, or GCS. The GCS specifies the other parameters necessary for latitude and longitude coordinates to be used with maximum accuracy when locating features and calculating distances and directions between them. These parameters include the measurement framework, its 3D unit of measurement, its prime meridian, and its reference ellipsoid and its datum.

1. Right-click the *World* data frame and select *Properties*
2. Click to select the *Coordinate System* tab
3. Scroll to the top of the top frame in the dialog box that appears

You should see that you are looking at information stored in the *Geographic Coordinate Systems* folder. Note that the information about your coordinate system for this exercise is stored in the *Geographic Coordinate Systems\World* folder. The details of that information are presented in the lower box.

QUESTION 3.4

What is the name of the geographic coordinate system used for this exercise?

    a.   GCS_WGS_1984

    b.   GCS_NAD_1983

    c.   GCS_Sphere

    d.   GCS_Clarke_1866

Exploring the Graticule's Geometric Properties

There are four geometric properties of the graticule that you should be able to recognize, so that you can assess how transforming 3D coordinates to a 2D map form alters them: *distance, shape or angle, area,* and *direction*. In this section, we will focus on distance and area.

It will help here if you can view the graticule as it would look on the globe:

1. *Right-click* on the *World* data frame and select *Properties* if you closed the window from the previous section
2. Click to select the *Coordinate System* tab
3. Scroll to the bottom of the top frame in the dialog box
4. Click on the + next to *Projected Coordinate Systems* to expand the folder
5. Scroll down and click on the + next to *World* to expand that folder
6. Scroll down and click once to select *Vertical Perspective (world)*
7. Right-click on *Vertical Perspective* and choose *Copy and Modify*
8. Select the value for *Latitude of Center* and change it to **45**
9. Click *OK*, then *OK* again
10. Zoom in to the intersection of the Equator and the Prime Meridian (a scale of around 1:20,000,000)
11. In the *scale window* in the toolbar, choose the 1:10,000,000 scale
12. In the *TOC*, click on *Graticule_1_Degree_Increments* to make it visible
13. Click to select the *Measure* tool in the toolbar
14. Click to select *Measure Line*
15. From the *Choose Units* drop-down arrow , set the *Distance* to *Miles*
16. From the *Choose Measurement Type* drop-down arrow, choose *Geodesic* to measure in 3D
17. Click on the equator line (**0°**), and then double-click on a line of latitude **1°** away

QUESTION 3.5

What is the approximate distance between these 2 lines of latitude? _____

QUESTION 3.6

Measure between 2 adjacent lines of longitude near the equator. Double check that your *Choose Measurement Type* is still set to *Geodesic*.

What is the approximate distance between these 2 lines of longitude? _____

QUESTION 3.7

1. Zoom out and pan to 60° latitude
2. Zoom in to that area of the globe to a scale of about 1:10,000,000 and re-measure the distance between 2 adjacent lines of latitude. Double check that your *Clear & Reset Results* is still set to *Geodesic*.

What is the approximate distance between these 2 lines of latitude? _____

QUESTION 3.8

Measure between 2 adjacent lines of longitude in the same area you measured for Question 3.7. What is the approximate distance between these 2 lines of longitude? _____

Mapping Earthquakes and Volcanoes

What types of spatial relationships exist between earthquakes and volcanoes? Mapping them is a good way to begin investigating the relationships between the two. A subset of earthquake and volcano locations is located in *Earthquake_Volcano*. This file is *not* a spatial data file; it is simply a listing of latitude and longitude values for several volcano and earthquake locations, saved in a spreadsheet format. To plot these locations in the display window, you must ask ArcMap to display them as (x,y) coordinate data.

1. Right-click *Earthquake_Volcano* in the *TOC* and choose *Open*

QUESTION 3.9

When the latitude value is specified as a negative number, then the point's location is north of the Equator.

a. True
b. False

QUESTION 3.10

When the longitude value is specified as a positive number, the point's location is:

a. north of the Equator
b. east of the Prime Meridian
c. west of the Prime Meridian
d. south of the Equator

CREATING GEOGRAPHIC FEATURES FROM COORDINATE LISTINGS

1. Right-click the table in the *TOC* and choose *Display XY Data*
2. In the dialog box that appears, your table should be listed, with the *Longitude* column in the *x field* and the *Latitude* column in the *y field*
3. Click the *Edit* button, and expand the *Geographic Coordinate Systems* folder

4. Expand the *World* folder and select WGS 1984
5. Click *OK*, then *OK* again if a Warning dialog box appears
6. Now, Right-click the *World* data frame and select *Properties*
7. Expand the *Geographic Coordinate Systems* folder
8. Expand the *World* folder and select WGS 1984
9. Click *OK*

VISUALIZING CATEGORIES OF POINT FEATURES

1. Right-click your *Earthquake_Volcano Events* layer and choose *Properties – Symbology* tab (Figure 3.3)

FIGURE 3.3

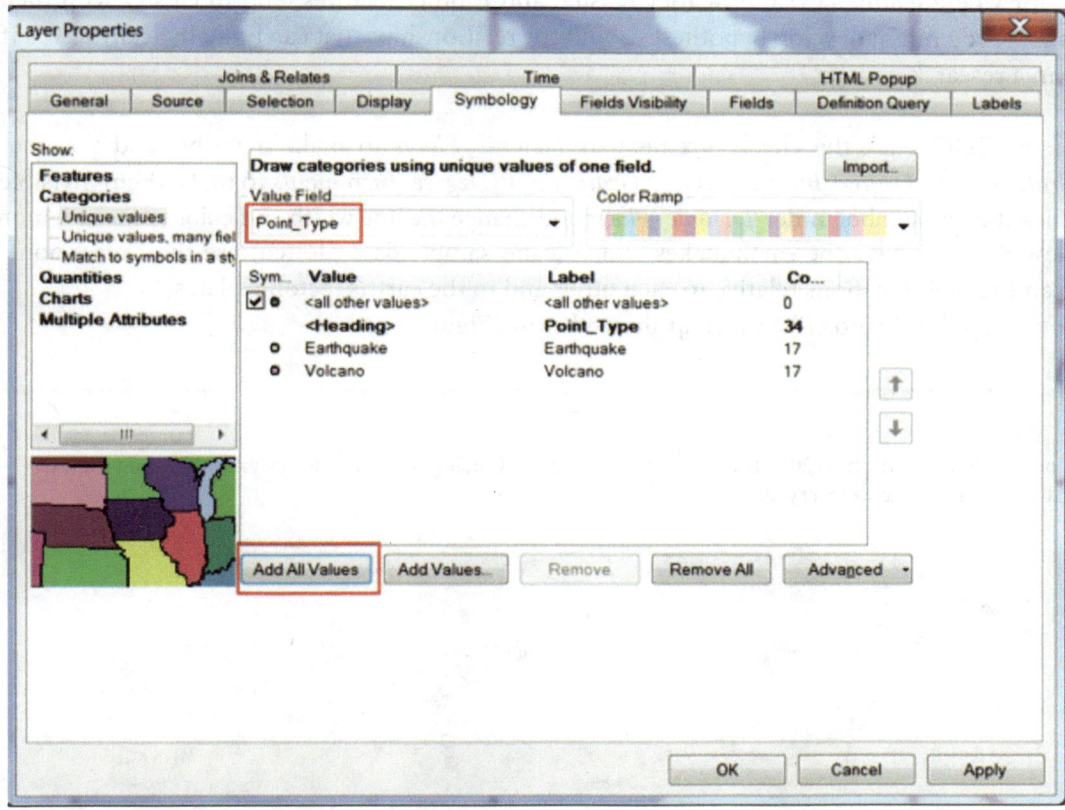

2. Under *Show:* on the left side of the dialog box, click to select *Categories*
3. In the *Value Field* that appears, choose the *Point_Type* field from the drop box
4. Click *Add All Values*
5. Double-click each symbol and choose symbols, symbol sizes, and symbol colors that are easily distinguished from one another, then click *OK*

Visualizing Relationships: Earthquakes, Volcanoes, & Plate Boundaries

Your map is a representation of the physical world, and is a great way to begin exploring the spatial relationships between earthquakes, volcanoes, and tectonic plate boundaries. Earthquakes and volcanoes are familiar features of our planet, but what is a plate boundary and why is it important?

Plate boundaries are part of the theory of plate tectonics, a concept from geology that explains most of our planet's major surface features and associated activities. According to plate tectonic theory, our planet's outer layer is composed of fragmented slabs of rock. These slabs overlay molten material that allows them to move relative to each other. The movements, in turn, create features like mountains, earthquakes, and volcanoes, to name a few. Most maps of tectonic plates show only the major plates; in reality, there are many smaller plates as well. If you would like to read more about this, try: http://pubs.usgs.gov/gip/dynamic/dynamic.html.

Maps like the one you have just created can be read, just like you might read a statistical graph. Typically, you look for where features exist, how they cluster, and if other features tend to cluster with them. From these insights, you might develop hypotheses about the relationships that can be statistically tested if you are working in a research context.

1. In the *TOC*, click the check box next to *Tectonic_Plates* to make it visible and the one next to *Graticule_20_Degree_Increments* and *Graticule_1_Degree_Increments* to make them invisible
2. Click the line symbol under *Tectonic_Plates* and change the line width and color to make it more visible
3. Now that you have the earthquakes and volcano sample data plotted, use the Data Zoom tools to examine their locations relative to each other and to the earth's tectonic plates.
4. Your map should look like the map shown below (Figure 3.4)

FIGURE 3.4

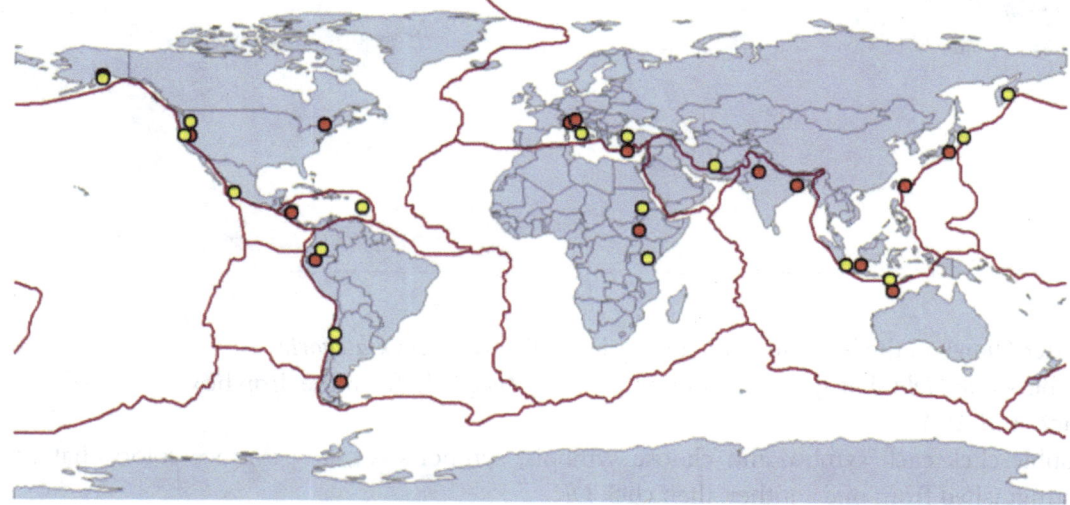

QUESTION 3.11

Our sample data includes earthquakes and volcanoes in the region of East Africa, where there is no apparent plate boundary within 1,000 miles of a volcano (hint: use the measure tool to estimate distances between volcano locations and the plate boundary along East Africa).

a. True
b. False

One result of plate tectonics is the *Ring of Fire*, an arc of volcanic and seismic activity that stretches around the edges of the Pacific Ocean (Figure 3.5). Similar in shape to a horseshoe, it arcs up from New Zealand into Japan and along the eastern edge of Asia, then across the Aleutian Islands of Alaska before heading south along the coasts of North and South America. This is an area where several tectonic plates meet the Pacific Plate.

FIGURE 3.5
Source: U.S. Geological Survey, Department of the Interior

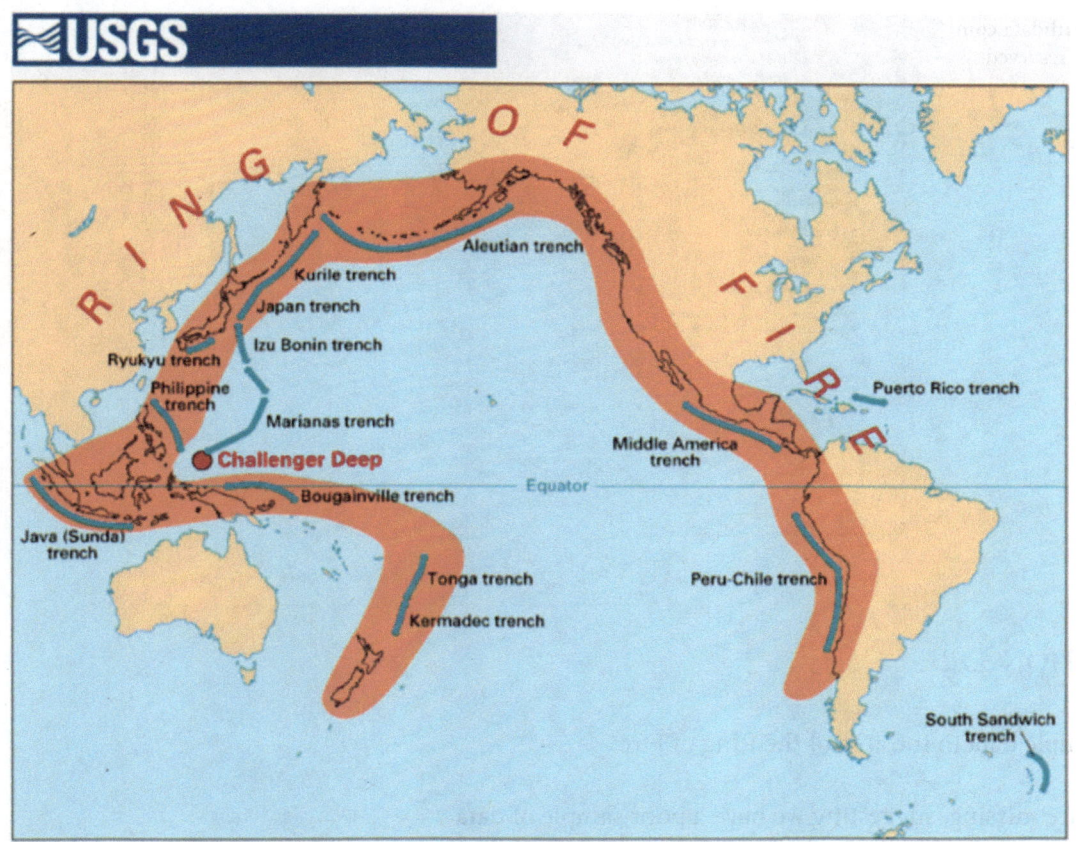

To center your view of the world on the Ring of Fire, complete the following steps:

1. Right-click the data frame and choose *Properties - Coordinate System* tab
2. In the top scroll box, scroll to find the *Projected Coordinate Systems* folder
3. Click the + sign to expand the *Projected Coordinate Systems* folder
4. Click the + sign to expand the *World* folder
5. Click to select *Robinson (world)*
6. Right-click *Robinson (world)* and choose *Copy and Modify*
7. Click in the *Central_Meridian value* box and type in the value of 180 to change the center of the display (Figure 3.6)
8. Click *OK*, then *OK* again.

FIGURE 3.6

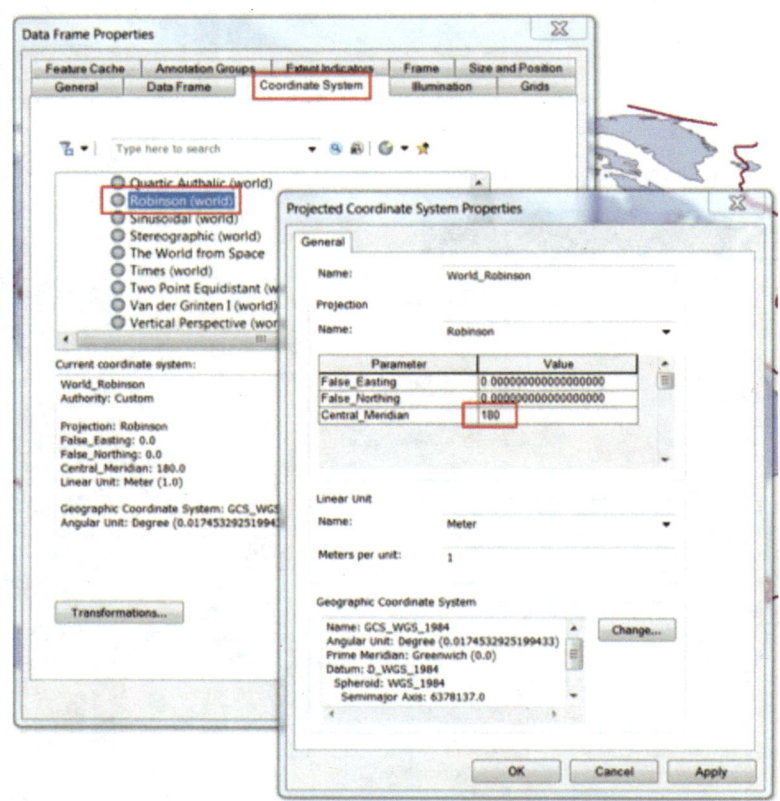

QUESTION 3.12

Our sample data in the area of the Ring of Fire:

a. are missing, suggesting we have a poor sample of data
b. do not align with the Ring of Fire, suggesting little correlation between plate boundaries and the location of earthquakes and volcanic activity
c. align well with the Ring of Fire, suggesting our data supports the link between plate boundaries and the location of earthquakes and volcanic activity

Export Your Work

1. Use the Data *Pan* tool to re-center your map within the Data View
2. Use the Data *Zoom* tools to make sure your entire map shows in the Data View
3. From the *File* menu, choose *Save*
4. From the *File* menu, choose *Export Map*
5. In the *Save In* drop-down box at the top of the dialog, navigate to the same location that you saved the original map document
6. Under the *Save as Type* drop-down box, choose *PDF* and then click *Save*
7. Navigate to the location where you saved your file and verify that it is there

Chapter 4

Projected Coordinate Systems

The geographic coordinate system pinpoints any location on the planet using its spherical system of latitude and longitude. It does this by creating a *graticule*, or network of intersecting meridians and parallels over the sphere. With this system in place, it was only natural to want to put it down on paper – a flat surface – to make it easier to carry while traveling. Systematically transferring these to a flat surface (a plane) using planar coordinates produces a *map projection*. When we transfer the locations from the sphere to a flat surface, however, we distort key properties of the original coordinate system. Because these properties are compromised, this affects how you might use any given map most accurately. To understand this process more clearly, let's examine one of the earlier map projections on record.

Moving from Sphere to Map

Maps, of course, existed in various forms long before the development of latitude and longitude. It was only as commerce with other nations began to increase that the need for more accurate maps soared. At this point, the use of latitude and longitude, along with accompanying methods of transforming this coordinate system to a flat map, became critical.

THE PLANE CHART

While Hipparchus provided us with the conceptual framework for the geographic coordinate system, it was a man by the name of Claudius Ptolemy who laid the groundwork for transforming these coordinates using map projections. Ptolemy (90–168 A.D.) published several writings on the subject; some of these were descriptive, while others, such as *Geographia*, outlined methods of mapping that would eventually contribute to the foundation of modern cartography (Brown, 1977). Even though the world was still awaiting the marine chronometer to measure longitude accurately, Ptolemy devised several projections using the *concepts* of latitude and longitude. He accomplished this by refining a nautical chart of the habitable world that a contemporary named *Marinus* had previously created around 120 A.D.

Marinus was a Greek geographer and mathematician. One of his contributions to geography was the development of a system of nautical charts that used latitude and longitude to map location over small areas. The result was a simple, rectangular, plane chart, known today as the *equirectangular projection* or, sometimes, the *Plate Carrèe projection* (Figure 4.1). The chart was composed of a grid of equidistant straight lines of latitude that were crossed at right angles by equidistant straight lines of longitude. It is a simple transformation, but, in the end, quite unsatisfactory. To understand why, however, you must compare the characteristics of the graticule on the sphere to what happens to those characteristics after Marinus transformed it to paper.

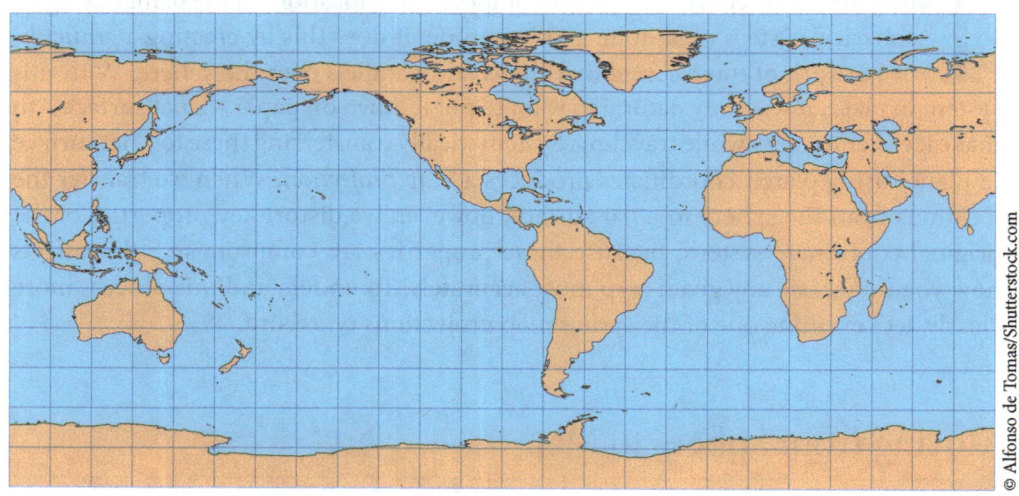

FIGURE 4.1
The Plane Chart, also known as the Equirectangular projection or the Plate Carrèe projection.

© Alfonso de Tomas/Shutterstock.com

CHARACTERISTICS OF THE GRATICULE

Four key geometric properties of the sphere – properties that we do not typically want our maps to distort relative to the original coordinate system of latitude and longitude – are **distances**, **shapes (local angles)**, **areas**, and **directions**. To recognize their distortion on a map, you must first recognize what these properties look like on the sphere. Dent and colleagues (2009) provide a good description of these properties; the following basics are derived from their discussion and are visualized in Figure 4.2:

1) Parallels and meridians have fixed distance relationships on the sphere. All meridians, for example, are the same length, and they are all equal to 1/2 the length of the equator. Parallels of latitude, on the other hand, decrease in length as you move from the equator to either pole, and this change can be modeled mathematically.

2) Meridians and parallels intersect at fixed angles. Parallels are always parallel to one another, and they are evenly spaced along meridians. Meridians are evenly spaced along parallels, but they converge at both poles and diverge as you move towards the equator. When a parallel and meridian intersects, it always forms a right angle.

FIGURE 4.2
Key geometric properties of the sphere. Data: Made with Natural Earth. Free vector and raster map data @ naturalearthdata.com. Source: Elisabeth Nelson

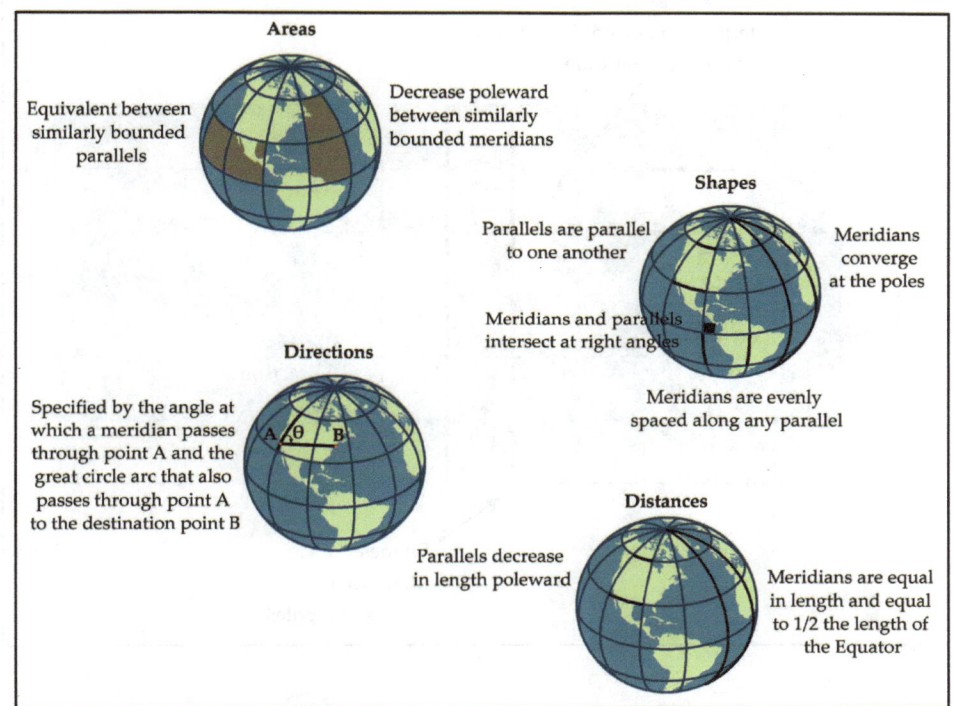

3) Intersecting parallels and meridians form quadrilaterals, which have the property of area. Areas are equal for all quadrilaterals that lie between two parallels, as long as they have the same longitudinal extent. On the other hand, areas of quadrilaterals between any two meridians vary with latitudinal extent.

4) Direction is specified by the angle formed by a meridian passing through a point and the great circle arc (the shortest distance between two locations on the sphere) also passing through that point to the destination point. This direction is traditionally reported as the angle clockwise from geographic north.

With these characteristics in mind, let's look at the graticule characteristics of the equirectangular or Plate Carrèe projection (Figure 4.3). First, since the lines of latitude are all equidistant, you should note that distances become stretched east to west as you move north and south of the equator. The lines of longitude are also straight and equidistant; distances along them, thus, remain correct, but the lack of convergence at the poles distorts the areas and shapes of quadrilaterals as you move poleward from the equator. General directions are also distorted in this transformation. In other words, this projection preserves none of the basic geometric properties over an entire map. Because mapmakers typically are looking to preserve one or more of these properties, this limits the usefulness of the projection for most mapping applications.

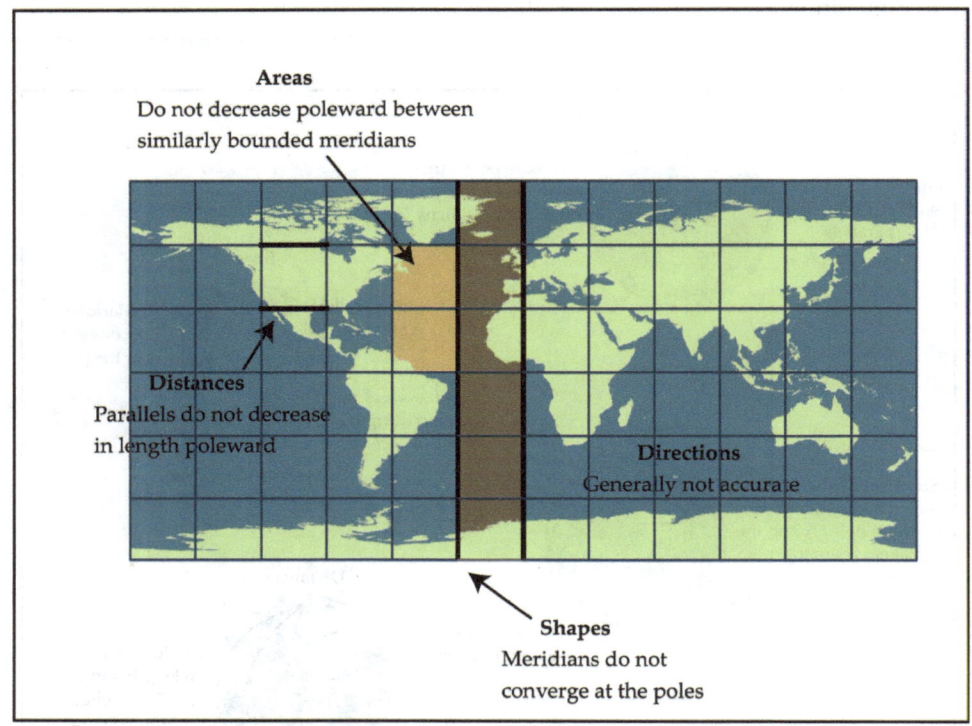

FIGURE 4.3
Distortion of key geometric properties: the Equirectangular or Plate Carrèe projection. Data: *Made with Natural Earth. Free vector and raster map data @ naturalearthdata. com.* Source: Elisabeth Nelson

TRANSFORMING DMS TO DD

Despite these limitations, many geographic information systems use the Plate Carrèe projection as a "default" projection when initially displaying spatial data. From a software perspective, this is not unusual. It is a simple projection to program, and the result is a familiar Cartesian grid. The grid uses the equator as the x-axis and the prime meridian as the y-axis, subdividing the world into four quadrants and using positive and negative signs in place of directional designations, a format more suitable to the computer world (Figure 4.4). Note that when projected into the digital realm, the familiar (latitude, longitude) description found in a typical atlas changes to (longitude, latitude); we determine our east-west position by moving along the x-axis, which, paradoxically, is the equator, a line of latitude. This transformation in coordinate specification moves us from the traditional degrees, minutes, seconds (DMS) to decimal degrees (DD). So, the Plate Carrèe projection finds a "home" for initial GIS computations, but it is critical for a GIS user to recognize the limitations of the projection from a map use perspective.

FIGURE 4.4
Latitude and longitude expressed as decimal degrees.
Data: *Made with Natural Earth. Free vector and raster map data @ naturalearthdata.com.*
Source: Elisabeth Nelson

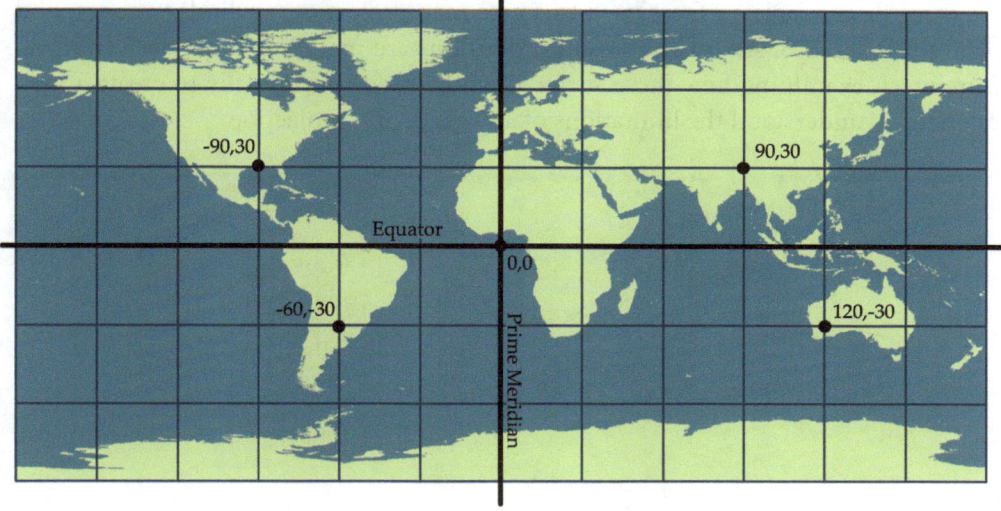

Classes of Map Projections

Marinus' plane chart may have been the dominant projection inherited from the Greek period, but Ptolemy also described others. He actually created two compromises of the Plate Carrèe that he believed retained spherical relationships more accurately. His first projection also had meridians drawn as straight lines, but instead of being parallel straight lines, they converged at the North Pole. From this projection, he then outlined another in which he curved the meridians. The curved meridians provided a better visual representation, but were much more difficult to construct by hand, so saw less use than the one with straight, converging meridians.

Ptolemy's *Geographia* is critical to further innovations not only for map projections, but also for general cartographic principles; it is the starting point for measuring all progress in western cartography (Thrower, 1996). With the fall of Rome, Europe entered a period in which theology overshadowed scientific interest, and works such as Ptolemy's, based heavily on mathematics, were largely forgotten or hidden. Because of this, his ideas weren't improved upon until *Geographia* was rediscovered by Europe at the beginning of the Renaissance in the 1300s. Many of the maps of the Middle Ages were more philosophical than scientific, and with the new push to explore the world, explorers and cartographers needed a mathematical basis to map coastlines more accurately. Ptolemy's work fit the bill nicely.

DEVELOPABLE SURFACES

Marinus' plane chart and Ptolemy's revised projections served as the prototypes for developing modern map projections. A *developable surface*, which is a surface that can be flattened into a plane without tearing, stretching or compressing that surface, forms the foundation for discussing the basics of projection development. There are three developable surfaces used to create map projections: cylinders, cones, and planes (Figure 4.5). Although current projections are mathematically derived using computers, examining how these surfaces can be used conceptually to create a map can help us understand the limitations of any particular projection.

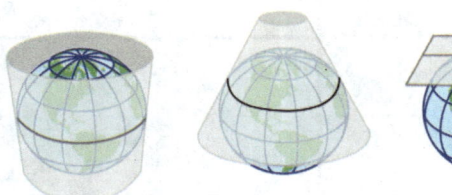

FIGURE 4.5
Developable surfaces and the reference globe. Data: *Made with Natural Earth. Free vector and raster map data @ naturalearthdata.com.* Source: Elisabeth Nelson

We begin with one of these surfaces and a *reference globe*, which is a ". . . reduced model of reality" (Dent, et al., 2009). The size of the reference globe depends on the size of the final map. Ideally, once you have the reference globe, you would simply peel the surface off and lay it flat to produce your map. Unfortunately, making a spherical surface lie flat without tearing, stretching, or compressing it is impossible (Figure 4.6), which leaves us to figure out how to transfer the features of the reference globe to either a cylinder, cone, or plane.

FIGURE 4.6
Making a spherical surface lie flat requires tearing, shearing, and compressing it.

© Anastasios Kandris/Shutterstock.com

Take Marinus' equirectangular projection, for example. This projection is a prototype of the cylindrical class of projections. In this instance, a cylinder is placed over the reference globe so that its long axis is parallel to the earth's axis. As the cylinder is placed over the globe, it touches or becomes *tangent* along the equator. If you were to imagine a light inside the globe projecting its graticule to the surface of the cylinder, you could trace the graticule, and then simply unroll the cylinder to lay the map flat. Cylindrical projections typically produce straight, equally spaced, parallel lines of longitude and straight, parallel, equidistant lines of latitude; the lines of latitude and longitude meet at right angles (Figure 4.7a).

FIGURE 4.7

Distortion patterns associated with developable surfaces projected using a normal aspect.
Data: *Made with Natural Earth. Free vector and raster map data @ naturalearthdata.com.*
Source: Elisabeth Nelson

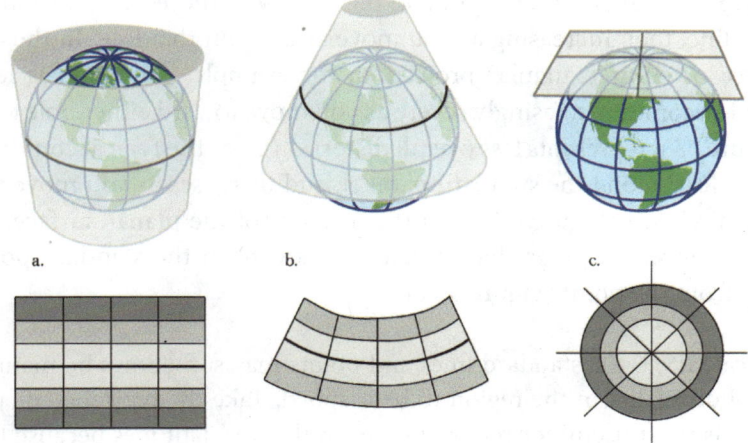

a.   b.   c.

While cartographers traditionally used cylinders to map the entire world, they found that cones and planes handled regions more accurately. Ptolemy's first compromise of Marinus' projection is an example of a simple conic projection (Figure 4.7b). A cone wrapped around the reference globe will become tangent to a mid-latitude parallel rather than the equator; while it cannot transfer information from the entire reference globe, it might be better relaying information about the middle latitude regions. Transferring the information needed using this surface, and then unrolling it produces a conic projection. The resulting transformation is wedge-like, where meridians are straight lines radiating from a central point, such as the North Pole, and lines of latitude are concentric arcs of small circles.

Our third surface, the plane, has actually been in existence as a projection longer than either the cylinder or cone, but the Greeks primarily used it to map the heavens, not our planet. Placing a plane against the reference globe means that the surface is tangent to a point, not a line. A common point of tangency in this instance is the North Pole. Like the cone, cartographers cannot use the plane to map the entire earth, but rather use it for mapping smaller regions of our planet. Planes also produce distinctive graticule patterns; here, the results are straight meridians radiating from the center of the projection, or standard point, and parallels that are equally spaced concentric circles centered on the standard point (Figure 4.7c).

Key Projection Characteristics

STANDARD POINTS AND LINES

Regardless of which developable surface the cartographer uses, one of the most important characteristics of the final projection is the line or point at which the projection surface touches the reference globe. Known as the standard line (or standard parallel, if it's a parallel) or standard point, these key features mark the line or point along the map where the transfer of information from reference globe to map is made without distorting geographic features. Map distortion patterns, then, are related to the standard line or point of the map, with the least distortion occurring around the line, then increasing as you move away from that line. In the case of the Plate Carrèe or equirectangular projection, for example, distortion is least at the equator and becomes increasingly severe as you move toward either pole (the lightest gray in Figure 4.7a represents less overall distortion). For the typical conic projection, distortion is least along the standard parallel, and increases as you move away from it in either direction (Figure 4.7b). In the instance of the planar surface, distortion increases in concentric rings that radiate outward from the standard point as you move away from the point (Figure 4.7c).

In the typical GIS, these standard lines and points may sometimes be manipulated to help control distortion in the region to be mapped. Take the typical conic projection. This surface is most useful for regional maps in the mid-latitudes because the surface contacts the globe along a standard parallel as opposed to the equator (Figure 4.7b). In many of the conic projections, the user can specify the exact parallel to better control the overall distortion for the region of interest. For example, if we were to map the United States using a conic projection, we might specify the standard parallel at 40° N, but if we wanted to map Brazil, a better choice for a standard parallel might be 15° S.

PROJECTION CASE

So far, we have confined ourselves to projections with only one standard line, but mathematically, you can create projections that have multiple standard lines. Many conic projections, for example, are based on two standard lines, which are formed when the developable surface is conceptually pushed "through" the reference globe (Figure 4.8). These projections are more complicated, of course, but their advantage is that they provide even tighter control over distortion. Whether you have one standard line or multiple ones defines the *case* of your projection. A projection with one standard line produces a tangent case; the secant case has more than one standard line. Although more complicated, secant cases are important as they are widely used in today's software packages. We'll look at some specific examples in the following section.

FIGURE 4.8

Conic secant case projection.
Data: *Made with Natural Earth. Free vector and raster map data @ naturalearthdata.com.*
Source: Elisabeth Nelson

ASPECT

Aspect refers to how the projection surface is positioned relative to the reference globe. When the surface's axis of symmetry aligns with the axis of the reference globe, the aspect of the projection is called *normal*; this class of aspect produces the simplest graticules (see Figure 4.7). For the cylinder, the equator, or equatorial aspect, is the normal aspect. A cone, on the other hand, uses a mid-latitude parallel to produce its normal aspect, while the normal aspect of a plane is the polar aspect. You can also specify a *transverse* aspect, which rotates the projection surface 90 degrees from normal. Both transverse and oblique aspects produce complex graticules and are used primarily to better center the projection surface over an area to be mapped (Figure 4.9).

FIGURE 4.9

Example of a transverse aspect, resulting in a complex graticule: Transverse Mercator projection. Image: *Courtesy of NASA.*

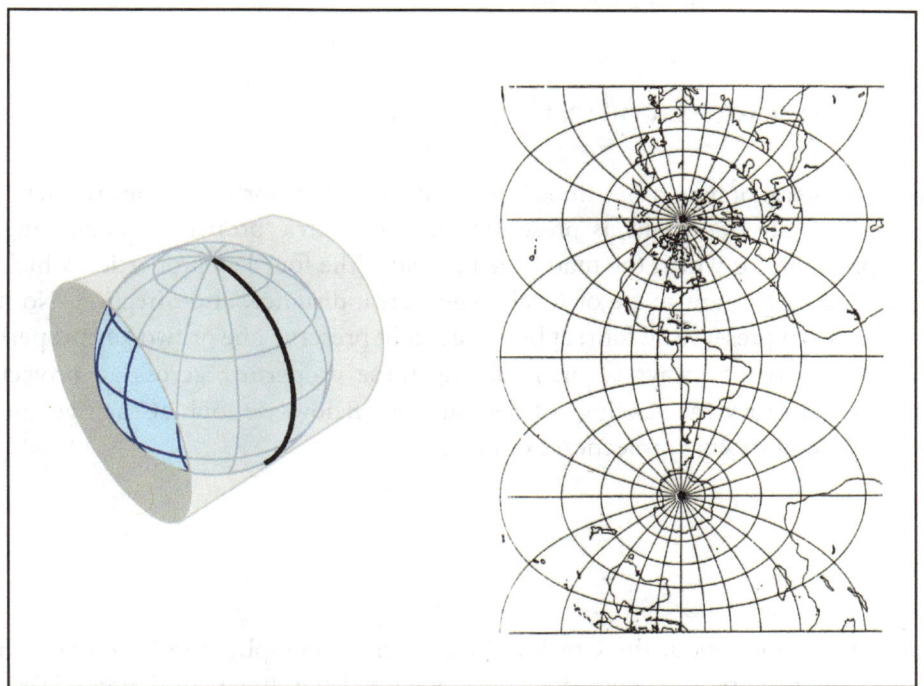

The concept of aspect categories is somewhat dated in the digital environment, where aspect is expressed and manipulated numerically (Strebe, 2009). In the world of GIScience, it is easier to think of aspect as a centering process, which involves precisely specifying a central meridian and standard parallels or points for a chosen projection (Figure 4.10). The central meridian is a specified meridian that is central to the area you are mapping. Specifying standard parallels as well helps minimize distortion within the mapped region. For example, if you were to make a map of the contiguous United States, you could minimize distortion by specifying a central meridian of 96° W and standard parallels at 20° N and 60° N.

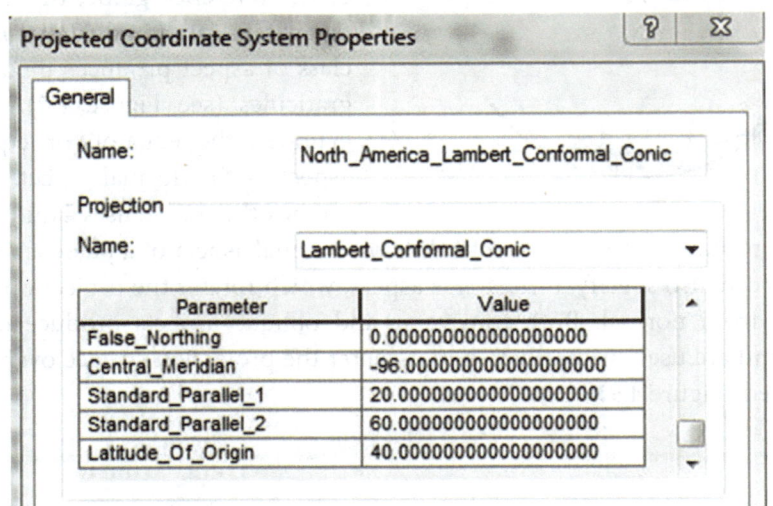

Preserving Key Geometric Properties

One of the goals of map projections is reducing overall distortion in the area mapped; another, equally important, is preserving key geometric properties, depending on the application for which the map is being made. The four key properties, which we mentioned earlier, are shape or local angles, area, distance, and direction. No map projection can preserve all four; at best, we might preserve one or two key properties. One of the primary ways of manipulating these properties, across all projection surfaces, is to vary the spacings of the parallels. Below, we look at each geometric property using specific projection examples.

SHAPE

We've talked a lot about the Greek influence on cartography, and have highlighted some of the prototypical projections developed by key scholars of that era. We have also noted that this period of scientific advancement ended with the fall of the Roman Empire and did not resurface until travel and exploration began to demand a need for more accurate map projections. At that point in time, navigators began to call for a map projection on which constant compass directions could be drawn with a straight line. Sailors had found it was one thing to sail within sight of land, and quite another to try to navigate over larger stretches of ocean, where land was not visible.

To create a map where lines of constant compass direction plot as straight lines requires preserving shapes or local angular relationships (property of *conformality*) across the map.

On the globe, meridians and parallels intersect at fixed angles, and direction is specified by the angle formed from a meridian passing through a point and the great circle arc also passing through that point to the destination point. Lines of constant compass direction, known as *loxodromes* or *rhumb lines*, then, are complex curves on the globe – they spiral towards the poles in order to cross each meridian at the same angle (Figure 4.11). To create a map on which rhumb lines plot as straight lines means preserving these angular relationships; the first person to achieve this was Gerardus Mercator, who produced one of the most famous projections of all – the *Mercator Projection* – in 1569.

FIGURE 4.11
Loxodrome or rhumb line as seen on a sphere.
Data: *Made with Natural Earth. Free vector and raster map data @ naturalearthdata.com.*
Source: Elisabeth Nelson

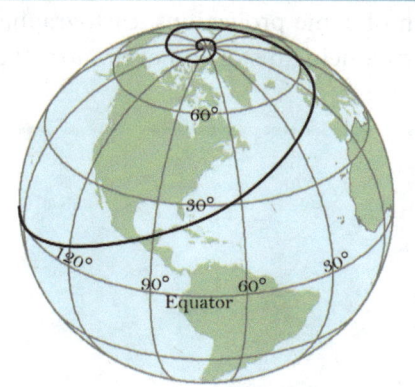

The Mercator projection (Figure 4.12) is a cylindrical, conformal projection that uses the equator as its standard line. On this projection, parallels and meridians intersect at right angles just as they do on the globe, providing our cardinal directions of north, south, east, and west. To compensate for angular distortions caused by non-intersecting meridians, Mercator graphically expanded the spacing of the parallels so that they became progressively further apart as the poles were approached. The rate at which he spaced them matched the rate at which the meridians were expanded. Today, we call this the *principle of equal stretching*; it corrects for the initial angular distortion caused by the projection, but greatly exaggerates distances and areas in doing so.

FIGURE 4.12
Mercator Projection.

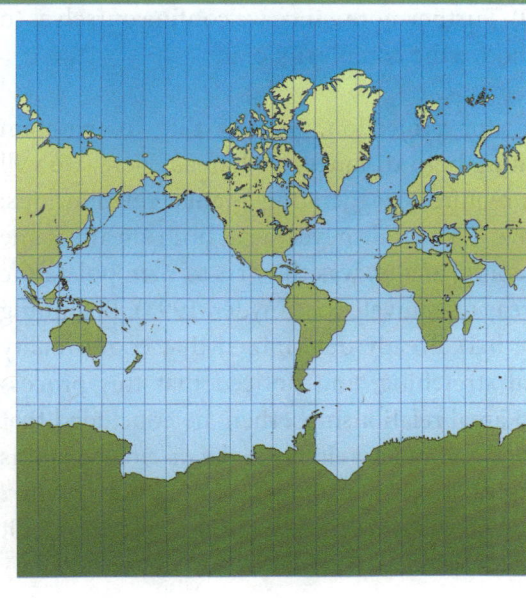

© Alfonso de Tomas/Shutterstock.com

While rhumb lines may now be drawn as straight lines, the use of this projection for other purposes is quite limited. To see what we mean, compare Greenland to the continent of South America. On the globe, Greenland is about 1/8 the size of South America, but on the Mercator projection, the relationship is different, with Greenland appearing to be much larger. In fact, as you approach the poles, distortion becomes progressively worse. At 60° N, for example, the parallel is ½ the length of the equator on the globe, but on the Mercator projection, the same parallel is twice the length. When you match that with an equal stretching of length between the meridians, the result is a fourfold increase in areas at that latitude. The Mercator projection, then, is most useful for navigation, and for mapping regions near its standard line, the equator (within about 35° of the equator), where distortion in geometric properties is minimal.

Two other conformal projections that you will run across quite frequently are the *Lambert Conformal Conic* and the *Transverse Mercator*, both invented by Johann Lambert in the late 1700s. Each was developed because the mapping community needed more precise projections on which to plot countrywide surveys and newly discovered coastlines. With the development of calculus in the field of mathematics, Lambert was able to harness new ideas that allowed for more precision in projection development. The Lambert Conformal Conic, for example, was designed to provide more accurate mappings for mid-latitude regions; not only was the standard line a mid-latitude parallel of choice, but with this revision of conic projections, cartographers could now specify two standard lines, effectively reducing distortion even more than a simple conic projection (Figure 4.13).

The Transverse Mercator was developed by conceptually rotating the cylindrical surface 90 degrees with respect to the globe. In using the transverse aspect, the long axis of the cylinder becomes perpendicular to the Earth's axis, and the surface is now tangent to the globe along a meridian of choice rather than the equator. While the world graticule is much more complex, this effectively makes the mappings of regions with

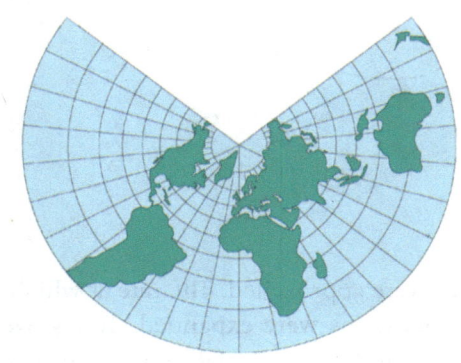

FIGURE 4.13
Lambert Conformal Conic Projection.
U.S. Geological Survey, Department of the Interior.

a long north-south extent more accurate than if we were forced to map them using a parallel as a standard line (see Figure 4.9). Furthermore, since we confine ourselves to a smaller part of the graticule, we don't notice the complexity of it.

Area. In our examination of the Mercator projection, we saw that to maintain angular relationships required exaggerating both distances and areas. Here, we will see that maintaining areal relationships requires exaggerating angular relationships; the two properties are *mutually exclusive* – you can never have both on the same map. Lambert, after having conquered conformality, was the first to manipulate the graticule deliberately to achieve equal areas or equivalence (Snyder, 1993). Beginning with a simple cylindrical projection, he had already determined that conformality was achieved by the principle of equal stretching, and noticed that this process deliberately distorted areas. To maintain areal relationships, then, he concluded that he would need to decrease the spacings of the parallels relative to the meridians of this projection (Figure 4.14). This narrowing of parallel spacings, creating the *Lambert Cylindrical Equal Area Projection*, is based on the sines of the latitudes, and the result is a distortion of angular, distance, and directional relationships to maintain areal relationships. This type of map, while not particularly useful for basic reference mapping, is the best choice for thematic maps, whose primary purpose is to compare spatial distributions between one or more regions.

FIGURE 4.14
Lambert Equal Area Cylindrical Projection. *U.S. Geological Survey, Department of the Interior.*

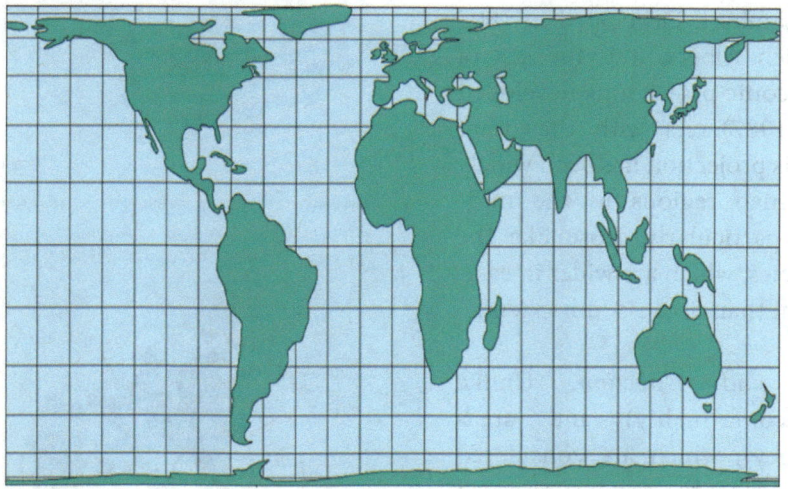

The *Lambert Azimuthal Equal Area Projection* and *Albers Conical Equal Area Projection* are two examples of equivalent projections that are commonly used in today's mapping. The Lambert Azimuthal Equal Area is an equivalent projection based on a plane surface (Figure 4.15). Producing a circular graticule centered on the North Pole, the parallels of latitude are concentric circular arcs that are centered on the pole, and the meridians are straight lines that intersect at the pole. Equivalency is preserved by spacing the lines of latitude closer together as you move away from the pole, similar to the spacings of the cylindrical equal-area projection. As you might expect, this projection is used to map polar areas; it is also useful, however, as a surface on which to map hemispheres, continents, and countries if you manipulate the aspect of the projection from normal (polar) to oblique or transverse.

FIGURE 4.15
Lambert Azimuthal Equal Area Projection. *U.S. Geological Survey, Department of the Interior.*

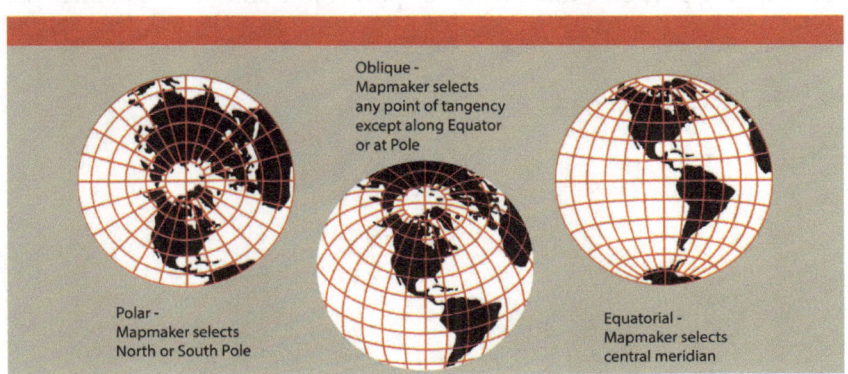

The Albers Conical Equal Area (Figure 4.16) is a special case of the Lambert Conical Equal Area, where the secant case of a conic surface is used, resulting in two standard parallels instead of the tangent case used by Lambert. Developed in 1805 by Heinrich Albers, it is the last of the basic conic projections developed (Snyder, 1987). As with the other conics, this projection has been widely used to map regions in the mid-latitudes, particularly those like the United States, which are wider in east-west extent than north-south extent.

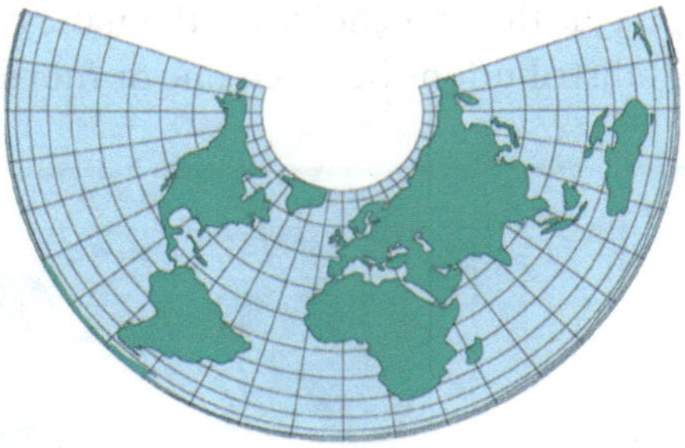

Distance and Direction. Unlike angles (conformality) and areas (equivalency), which are considered global properties that can be preserved over an entire map, preservation of distances and directions are limited to only certain areas of a map due to the transformation process. Because of this, these two properties are being considered together as local properties. Examples of projections that you might run across or want to use that preserve these properties are the Lambert Azimuthal Equal Area Projection (Figure 4.15) and the Azimuthal Equidistant Projection (Figure 4.17). Both of these are developed on a plane surface – we just discussed the equal-area property of Lambert's projection above – and both preserve directions from the standard point to all other points on the map. The Azimuthal Equidistant also preserves distances from the center point to all other points, making it useful for mapping purposes that involve airport traffic and missile launch sites.

FIGURE 4.16
Albers Equal Area Conic Projection. *U.S. Geological Survey, Department of the Interior.*

FIGURE 4.17
Azimuthal Equidistant Projection. *U.S. Geological Survey, Department of the Interior.*

Additional Readings

American Cartographic Association. 1988. *Choosing a World Map.* American Congress on Surveying and Mapping: Falls Church, Virginia.

American Cartographic Association. 1991. *Matching the Map Projection to the Need.* American Congress on Surveying and Mapping: Bethesda, Maryland.

Brown, L.A. 1977. *The Story of Maps.* Dover Publications, Inc.: New York, New York.

Dent, B.D., Torguson, J.S., and T.W. Hodler. 2009. *Cartography: Thematic Map Design.* 6th ed. McGraw-Hill: New York, New York.

Robinson, A.H., Morrison, J.L., Muehrcke, P.C., Kimerling, A.J., and S.C. Guptill. 1995. *Elements of Cartography.* 6th ed. John Wiley & Sons, Inc.: New York, New York.

Slocum, T.A., McMaster, R.B., Kessler, F.C., and H.H. Howard. 2005. *Thematic Cartography and Geographic Visualization.* 2nd ed. Pearson Education, Inc.: Upper Saddle River, New Jersey.

Snyder, J. 1987. *Map Projections: A Working Manual.* Washington, D.C.: U.S. Geological Survey Professional Paper 1395.

Snyder, J. 1993. *Flattening the Earth.* University of Chicago Press: Chicago, Illinois.

Strebe, D. 2009. *Map Projection Essentials.* Retrieved from http://www.mapthematics.com/Essentials.php, August 2010.

Thrower, N.J.N. 1996. *Maps and Civilization: Cartography in Culture and Society.* University of Chicago Press: Chicago, Illinois.

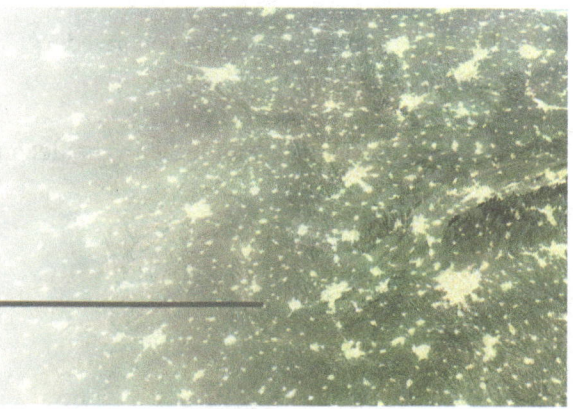

Exercise 4

Map Projection Basics

In Exercise 4, you will examine the foundations of map projections, manipulate their characteristics, and explore several specific projections that are popular for different applications

OBJECTIVES

- Examine the default parameters the GIS uses to display data
- Set a projected coordinate system
- Examine the differences between the projected coordinate system and the default display
- Explore projection classes, characteristics, and distortions

SOFTWARE INFORMATION

Introducing ArcMap uses Esri's **ArcGIS 10.6** software. **ArcGIS** is a commercial GIS package, available in geospatial labs affiliated with schools that have a campus-wide site license for the software. Instructors at these campuses may also request 1-year student versions of the software at http://www.esri.com/landing-pages/education-promo.

DATA

The data for the exercises is available from the Kendall Hunt Student Ancillary site. See the inside front cover for access information. You may also be directed to download the data from a different location by your instructor.

1. Download the data as instructed by your instructor
2. Save the file to a location where you have read/write privileges (USB key, home computer, class server)

The file you just saved is in a compressed (*.zip) format, and was created using **7-zip** freeware (http://www.7- zip.org). To use the data for this exercise, you must decompress it using the same software.

3. Locate the zipped file that you saved
4. Right-click the file
5. Choose *7-Zip - Extract Files*
6. Click *OK* to create a folder with the decompressed data

The Default Display in ArcMap

CONCEPT

One of the earliest projections on record is the equirectangular, or *Plate Carrèe*, projection. This projection produces a simple, rectangular chart composed of a grid of equidistant straight lines of latitude that are crossed at right angles by equidistant straight lines of longitude.

It is a simple transformation, but the resulting map preserves none of the basic geometric properties most cartographers require. Despite these limitations, many GIS use a display similar to the Plate Carrèe for the initial visualization of spatial data. Not only is it easy to implement, it also stores the spatial coordinates in their original 3D form, which enhances analytical accuracy.

1. Start **ArcMap**. You can access ArcMap using several different methods, depending on your environment. You can start by clicking the Windows start menu icon at the bottom left corner of your screen, then navigate to the ArcMap icon in the ArcGIS folder, or search for ArcMap in the search window by typing "arcmap".
2. In the *ArcMap - Getting Started* dialog box, click *Cancel*
3. If the *Catalog* window is not open, click on the *Catalog* icon on the Standard toolbar, or the *Catalog* tab on the right side of the **ArcMap** interface to open it
4. Click the *Pushpin* in the upper right corner of the *Catalog* window to freeze the window in place
5. Right-click *Folder Connections* and choose *Connect to Folder*
6. Navigate to the location where you saved the *Chapter04* folder of data for this exercise
7. Click once to select the *Chapter04* folder, then click *OK* to create a folder connection that will allow you to access the data inside the folder
8. Click the + sign to the left of the folder to see its contents. You should see a geodatabase for *Exercise_04*
9. Click the + sign to the left of the geodatabase to see the actual spatial data files you will use in this exercise
10. Click and drag each data layer from the *Catalog* window to your *Data View* window
11. In the *Table of Contents* (TOC) window, click on the *List by Drawing Order* icon
12. Click and drag the data layers so that the *Tissot_Indicatrix* layer is above the *Graticule_15_Degree_Increments* layer and both of these are above the *World_Countries* layer
13. Uncheck the *Tissot_Indicatrix* layer to make it invisible
14. Double-click the *Layers* data frame in the *TOC*
15. Click the *Coordinate System* tab to select it

QUESTION 4.1

What coordinate system is listed for this data frame?

a. Sphere_Mercator
b. GCS_North_American_1983
c. GCS_WGS_1984
d. World_Plate_Carrèe

QUESTION 4.2

Is the current coordinate system a *geographic coordinate system* only or does it also have a *projected coordinate system* as part of its spatial reference? _____

16. Click *OK*
17. Right-click the *Graticule_15_Degree_Increments* layer and choose *Properties*
18. Under the *Labels* tab, change *Label Field:* to **degrees** and click *OK*
19. Right-click the graticule layer again and choose *Label Features*

FIGURE 4.1

20. Zoom to *North America* using the *Data* zoom tools (Figure 4.1)
21. Set the scale in the *Map Scale* window to **1:15,000,000**
22. Use the *Pan* tool to navigate to 60° N and between 75° and 90° W
23. From the *Tools* toolbar, click on the *Measure Tool*
24. Under the *Choose Units* drop-down menu, choose *Distance - Miles*
25. From the *Choose Measurement Type* drop-down button, choose *Geodesic* to measure in 3D
26. Measure the longitudinal distance between the **75°** and **90° W** graticule lines along the **60° N** latitudinal line

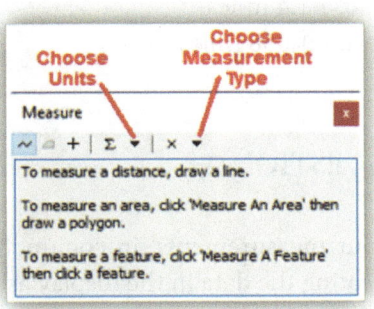

QUESTION 4.3

Which of the following measurements most closely approximates the distance between the two meridians at 60° N?

    a.   519 miles
    b.   52 miles
    c.   2,500 miles
    d.   5.24 miles

27.   Repeat this process to measure the longitudinal distance between the **75°** and **90° W** graticule lines along the **30° north** latitudinal line & the **0° N** latitudinal line

QUESTION 4.4

Which of the following measurements most closely approximates the distance between the two meridians at 30° N?

    a.   9 miles
    b.   380.8 miles
    c.   899 miles
    d.   89 miles

QUESTION 4.5

Which of the following measurements most closely approximates the distance between the two meridians along 0° N?

    a.   103 miles
    b.   34.8 miles
    c.   1.4 miles
    d.   1,038 miles

QUESTION 4.6

Your measurements, in conjunction with the stated coordinate system, provide evidence that **ArcMap** is treating the data in the display as 2D projected map data.

    a.   True
    b.   False

28. Close the *Measure* Tool window
29. Select the *Select Elements* tool

Save Your Map Document

1. From the *File* menu, choose *Map Document Properties*
2. Click the file folder button to the right of the default geodatabase path and browse to the *Exercise_04.gdb*
3. Click to select it, then click *Add*
4. Next to pathnames, place a checkmark by *Store relative pathnames to data sources*
5. Click *OK*
6. From *File*, choose *Save As*
7. Name your file using this convention: *lastname_Exercise04*
8. Save your map to the *Chapter04* folder containing *Exercise_04.gdb*
9. Double-click the *Computer* icon on your desktop
10. Navigate to the location where you think you saved your file and verify that it is there

Setting a Projected Coordinate System

CONCEPT

Setting a formal projection for your data will always distort your map's geometric properties, both visually and numerically. The specific projection you choose will dictate how each geometric property is specifically distorted, so different projections have different uses.

1. Right-click the **Layers** data frame, then click *Properties*
2. Click the *Coordinate System* tab
3. In the top scroll box, scroll to find the *Projected Coordinate Systems* folder
4. Click the + sign next to the folder to expand it and view its contents
5. Find the *World* folder and expand its contents
6. Find and select **Plate Carrèe**, then click *OK*
7. From the *Tools* toolbar, click on the *Measure Tool*
8. Under the *Choose Units* drop-down menu, choose *Distance - Miles*
9. From the *Choose Measurement Type* drop-down button, choose *Planar* to measure in 2D
10. Repeat the distance measurements between the 75th and 90th W longitude lines you made in the last section

QUESTION 4.7

The distance between 75° W and 90° W at 60° N is now approximately the same as the distance at 30° N and 0° N.

a. True
b. False

11. Close the *Measure* window and select the *Select Elements* tool

Classes of Map Projections

CONCEPT

While all map projections are mathematical transformations, the foundation of many of them is a *developable surface*, or geometric shape, that can be flattened without stretching the surface onto which spatial features are mapped. Common surfaces include the *cylinder*, *cone*, and *plane*.

The Cylinder

The *Plate Carrèe*, which was a prototype for developing modern map projections, is based on the cylindrical model. With this model, a cylinder is wrapped (theoretically) around the globe, features are transferred from the globe to its surface, and then the surface is unwrapped and laid flat, producing a map.

1. Right-click the **Layers** data frame, then choose *Properties* and the *General* tab
2. Rename this data frame **Cylindrical** and click *OK*
3. Click the *Zoom to full extent* icon to see the entire projection

QUESTION 4.8

In a simple cylindrical projection, like the *Plate Carrèe* or the *Mercator*, which of the following is NOT TRUE regarding the graticule's appearance?

a. the lines of longitude are straight lines
b. the lines of longitude are equally spaced along the lines of latitude
c. the lines of longitude are drawn at right angles to the lines of latitude
d. the lines of longitude converge at the Poles

The Cone

One of the first modifications proposed for the *Plate Carrèe* projection was to make the meridians behave more like they do on the globe. Using a conic surface forces the meridians to converge toward one pole because the cone, when wrapped around the globe, sits on the globe like a hat. When features are transferred to the cone, and then the surface is unwrapped and laid flat, the result is a map with a distinctly different graticule from the cylindrical model.

4. Click on the *Layout View* button in the bottom left-hand corner of your data view window
5. Click on the map within the data frame
6. From the *Edit* Menu bar, click *Copy*
7. From the *Edit* Menu bar, click *Paste*
8. Click on the *Data View* button in the bottom left-hand corner of your data view window
9. In the *TOC*, double-click the new data frame, click the *General* tab and change the name to **Conic**
10. Click *Apply*, then click on the *Coordinate System* tab
11. In the top scroll box, scroll to find the *Projected Coordinate Systems* folder
12. Click the + sign next to the folder to expand it and view its contents
13. Click the + sign next to *Continental* to expand that folder, then the + sign next to *North America* to expand it
14. Scroll down to find and select **WGS84 Canada Atlas LCC**
15. Click *OK*
16. Click the *Zoom to full extent* icon to see the entire projection
17. From the *File* menu, choose *Save* to save an updated version of your map document

QUESTION 4.9

Which of the following best depicts the appearance of the simple conic graticule?

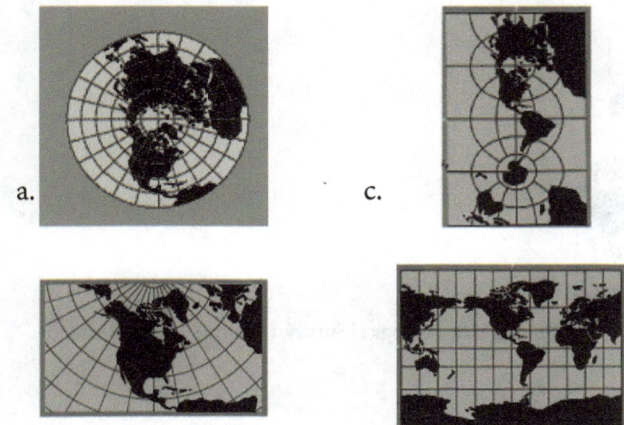

a.

c.

b.

d.

Source: U.S. Geological Survey, Department of the Interior

The Plane

The third developable surface is the plane. Unlike the cylinder and the cone, the plane is not wrapped around the globe but pinned to it at a point, usually a pole. When features are transferred from the globe to the plane, a third unique graticule is created.

18. Click on the *Layout View* button in the bottom left-hand corner of your data view window
19. Click on either map within the data frame
20. From the *Edit* Menu bar, click *Copy*
21. From the *Edit* Menu bar, click *Paste*
22. Click on the *Data View* button in the bottom left-hand corner of your data view window
23. Double-click the new data frame, click the *General* tab and change the name to **Plane**
24. Click *Apply,* then click on the *Coordinate System* tab
25. In the top scroll box, scroll to find the *Projected Coordinate Systems* folder
26. Click the + sign next to the folder to expand it and view its contents
27. Click the + sign next to *Polar* to expand that folder
28. Scroll down to find and select **North Pole Lambert Azimuthal Equal Area**
29. Click *OK*

Helpful hint: To switch back and forth between data frames, right-click the data frame you want to see and choose 'Activate'

QUESTION 4.10

Which of the following best depicts the appearance of the simple planar graticule?

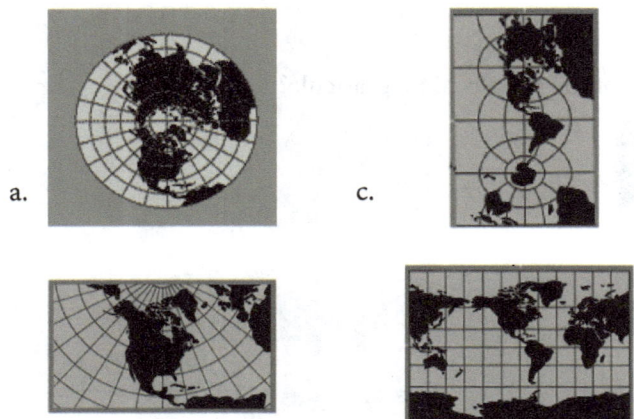

a.

c.

b.

d.

Source: U.S. Geological Survey, Department of the Interior

Projection Distortion

CONCEPT

The projection surface, in conjunction with other projection characteristics, will influence how the geometric properties of area, shape, distance, and direction are specifically distorted. The *Tissot_Indicatrix* layer consists of a set of *ellipses of distortion* that can be used to visually characterize projection distortion. Each ellipse describes the distortion occurring at a single point on the map, and is the result of projecting a circle of infinitesimal radius from a globe onto the map. By studying how each circle is deformed, distortions in feature areas and shapes, as well as the distances between features can be described:

- The smallest circles indicate areas on the projection that have minimal distortion
- Circles that change size indicate a deformation in the areal extent of a feature
- Circles that change shape indicate a deformation in feature shapes
- Circles in which the semi-major and/or semi-minor axes change length in relation to the smallest circles indicate a distortion in distance
- Directions cannot be evaluated with this tool

1. In the *TOC*, right-click the *Cylindrical* data frame and choose *Activate*
2. Click the box next to *Tissot_Indicatrix* to make the ellipses of distortion visible
3. Right-click the *Tissot_Indicatrix* layer and choose *Selection - Make this the only selectable layer*
4. Using the *Select Features* tool from the *Tools* toolbar, select a circle lying along the equator (Figure 4.2)
5. Right-click the *Tissot_Indicatrix* layer in the *TOC* and choose *Open Attribute Table*
6. Scroll to find the selected circle—note its *shape_area* and *shape_length* attributes—the shape_length attributes are the lengths of the semi-major and semi-minor axes
7. Back in the *Data View*, select another circle along the 60° N latitudinal line
8. In the attribute table, note its *shape_area* and *shape_lengths*

FIGURE 4.2

Data: *Made with Natural Earth. Free vector and raster map data @ naturalearthdata.com.*
Source: Elisabeth Nelson

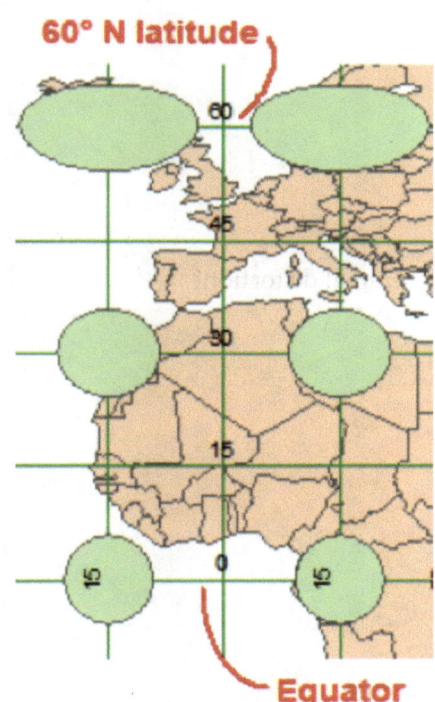

9. Using what you know about changes in the Tissot Indicatrix when map projections are modified (re-read the beginning of this section for more), answer the following questions:

QUESTION 4.11

In this projection (the Plate Carrèe), the shapes of map features are distorted.

   a. True
   b. False

QUESTION 4.12

In this projection (the Plate Carrèe), the areas of map features are distorted.

   a. True
   b. False

QUESTION 4.13

In this projection (the Plate Carrèe), general directions are distorted, except locally along the standard parallels.

   a. True
   b. False

QUESTION 4.14

Along which line of latitude is there minimal distortion?

   a. 90° N
   b. 60° N
   c. 30° N
   d. 0° (Equator)

9. Close the attribute table

Projection Characteristics

CONCEPT

Standard Lines

Regardless of which developable surface the cartographer uses, the first step in projecting features from the reference globe to that surface is to establish either a *point* or *line of contact* between the two models. These contacts, where the surface touches the globe during the transfer of features, are known as *standard lines* and *standard points,* or *lines* and *points of tangency.* Like the smallest circles plotted when using a Tissot's Indicatrix of distortion, this characteristic is significant because it marks locations on the map where there is minimal distortion in the transfer of geographic features.

1. On the Main Menu, click *Help*, then *ArcGIS Desktop Help*
2. Click the *Contents* tab, and then the *Search* tab at the top of the *TOC*
3. Type *Plate Carrèe*, and press *Enter*
4. Click the top result under *Click to view*, which should be *Plate Carrèe*
5. Scroll through the Help information to answer the following questions

QUESTION 4.15

What is the standard line or line of contact for the *Plate Carrèe* projection?

a. 45° N
b. 0° (Equator)
c. 60° N
d. 90° N (North Pole)

QUESTION 4.16

According to *Help*, the geometric properties of *shape* and *area* are preserved with this projection.

a. True
b. False

6. Close the Help window
7. Right-click the *Conic* data frame and choose *Properties,* then *Coordinate System* tab

QUESTION 4.17

What are the standard lines or standard parallels for this *Lambert Conformal Conic* projection?

    a.   0° N
    b.   90° N and 90° S
    c.   49° N and 77° N
    d.   29° N and 45° N

QUESTION 4.18

Given the choices of the *Plate Carrèe* or the *Lambert Conformal Conic* projection used in this exercise, which choice would offer the least amount of overall distortion for mapping the city of Nairobi in the country of Kenya, where Kenya spans approximately 5° N to 5° S latitudinally and 34° E to 42° E longitudinally?

    a.   Plate Carrèe
    b.   Lambert Conformal Conic
    c.   Either would work equally well

Centering

Overall distortion is also minimized by *centering* the projection. When you center a projection, you customize its standard parallel(s) and central meridian to correlate with the area of the world you are mapping. We could, for instance, use the *WGS 1984 Canada Atlas LCC* to make a map of Brazil if we re-centered the projection.

8.   Click *Cancel* to close the *Data Frame Properties* dialog box
9.   Right-click the *Conic data* frame and choose *Activate* to view it in the *Data View*
10. Check the box next to *Tissot_Indicatrix* view the distortion ellipses
11. Take note of the current patterns of distortion, particularly around Brazil (located approximately 5° N to 33° S latitudinally and 34° W to 74° W longitudinally)
12. In the *TOC*, right-click the *Conic* data frame and choose *Properties*
13. Click the *Coordinate System* tab, then scroll to find the highlighted projection
14. Right-click the projection and choose *Copy and Modify*, then change the *Central_Meridian* to **-52**, the *Standard_Parallel_1* to **0**, & the *Standard_Parallel_2* to **-30** (Figure 4.3)
15. Change the name from *WGS_1984_Canada_Atlas_LCC* to *WGS 1984 Brazil LCC* and click *OK*, then *OK* again
16. Click the *Zoom to Full Extent* icon

FIGURE 4.3

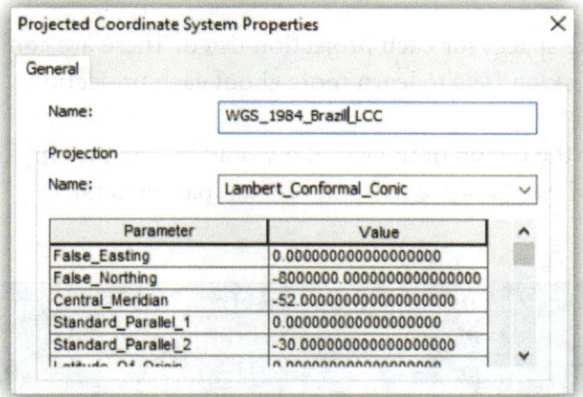

QUESTION 4.19

When mapped using *WGS_1984_Canada_Atlas_LCC*, map distortion is lowest in the latitudinal belt that includes Canada and increases as you move south towards the South Pole. When you modify the projection to center it on Brazil, the most distorted circles plot in:

a. Kenya
b. Canada
c. Mexico
d. South Africa

Preserving Geometric Properties

While reducing overall distortion for a mapped area is a key component driving the choice of a projection, equally important is preserving one or more key geometric properties. No map projection can preserve all four; at best, we might preserve one or two for any given application.

Shape. To create a map for traveling or for use as a basic reference that might be used to estimate distance and direction over a limited area requires preserving the shapes of local features. This property is called *conformality* and projections that preserve conformality are known as *conformal projections.*

Area. Preserving the areas of different features relative to one another is not particularly useful for navigation, but it is the best choice for thematic maps, whose primary purpose is to compare spatial distributions between one or more regions. This property is called *equivalency* and projections that preserve area are known as *equivalent* or *equal-area* projections.

Distance & Direction. Unlike conformality and equivalency, which can be held across an entire map, preserving distances and directions is only possible for certain areas of a map. Typically, both properties are preserved only from the map's standard point to all other points on the map. Projections that preserve distance are termed *equidistant*, and projections preserving directions are called *azimuthal*.

QUESTION 4.20

For the chart below, fill in the spaces for each projection listed. These are commonly used projections; you can use the *Help – ArcGIS Desktop Help* to learn more about each projection.

For the last column, indicate the region of the world it would be most useful for mapping from this list: *Equatorial, Mid-latitudes,* or *Poles* (assume use of its default parameters).

Projection	Developable Surface	Standard Line(s)/ Point(s)	Properties Preserved	World Region Use
Albers Equal Area				
Azimuthal Equidistant				
Lambert Azimuthal				
Lambert Conformal				
Miller				
Mollweide				
Stereographic				

17. Use the *Zoom to Full Extent* tool to zoom out so your entire map shows within the Data View
18. From the *File* menu, choose *Save*
19. Now, from the *File* menu, choose *Export Map*
20. In the *Save In* drop-down box at the top of the dialog, navigate to the same location that you saved the original map document.
21. Under the *Save as Type* drop-down box, choose PDF
22. Name your file to match the name of the original map document, then click *Save*
23. Double-click the *Computer* icon your desktop
24. Navigate to the location where you think you saved your file and verify that it is there

Note: To receive full credit during the grading process, the map you export should reflect the latest changes as directed by the exercise.

25. Exit **ArcMap**
26. If you are on a machine in a lab and are finished with your work for this session, log off and/or restart your machine before you leave.

Chapter 5

Mapping Location

As the race for developing an accurate navigational timepiece raged, cartographers were also calling for improvement of land surveys. Governments wanted better documentation of the lay of the land within their borders; although earlier surveys of relatively small land areas existed, the standards used in creating them were not rigorous enough to map larger areas at the same scale. Information on such topics as elevation change and river flows, in particular, was sorely needed. The first country to respond to this call was France, where the Académie Royale des Sciences was formed in 1666. This society consisted of a distinguished group of scientists whose purpose was to inspire and improve scientific research in France. Funded by King Louis XIV, one of their first missions was to increase the accuracy of French land surveys. They began work when the Académie brought Jean-Dominique Cassini, an astronomer from Italy, onboard in 1669 (Brown, 1977).

Earth Models and the Survey of France

RE-ESTABLISHING THE EARTH'S CIRCUMFERENCE

After much discussion, the Académie's scientists decided the first step in increasing the accuracy of these survey maps would be to re-determine the length of one degree of latitude. This meant beginning the survey by re-establishing the circumference of the Earth. Ultimately, they decided to use a variant of Eratosthene's method (see Chapter 2), with the primary difference being that the French had access to better measuring methods and equipment to carry out the task (Brown, 1977). Jean Picard, a French astronomer, directed this aspect of the survey, with Cassini overseeing the larger operation. *Triangulation* was Picard's method of choice for solving the problem because it replaced the need to measure the distances between every pair of locations comprising the survey. Instead, his team calculated one distance measurement between two known locations to form a baseline, and then they applied trigonometric principles and angle measurements from each end of that baseline to a third location to determine the distances to that point (Figure 5.1). From this first triangle, the other two sides were then used as baselines to extend the network to other locations.

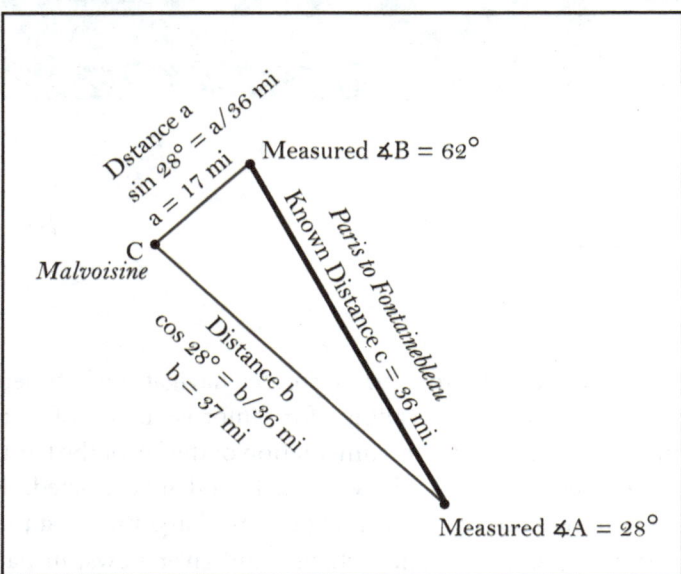

FIGURE 5.1
Triangulation Example.
Source: Elisabeth Nelson

The distance of the final meridian line, which ran north from Paris, was approximately 77.8 miles in modern-day units of measurement. With the length of that line established, Picard's surveyors then made astronomical observations at both of its endpoints to determine the length of the line in degrees, with the result being approximately 1° 11' 57" (1.2°). Multiplying the 77.8 miles by the number of 1.2° segments it takes to produce 360° (300), gives us an estimated circumference of 23,340 miles for our planet, which is quite close to today's measurement of 24,906 miles (Slocum, et al., 2005). Finished in 1670, Picard's meridian line, along with a topographic survey of Paris, formed the foundation for what would soon be the first large-scale survey of an entire country (Brown, 1977).

VARYING LATITUDE MEASUREMENTS CREATE CONTROVERSY

Following this initial survey and Picard's death in 1682, Jacques Cassini, son of Jean Dominique Cassini, took over and extended Picard's work. He surveyed one arc northward to Dunkirk, and another southward towards the boundary with Spain. When he calculated the length of a degree of latitude in both arcs, however, he discovered a difference, which cast doubt on the accepted convention of the Earth being spherical in shape. By his calculations, the length of a degree in the northern arc was shorter than the one in the southern arc, suggesting that the Earth might actually be a *prolate spheroid* (Figure 5.2), or, in more descriptive terms, egg-shaped (O'Connor and Robertson, 2008).

FIGURE 5.2

Proposed shapes for Earth: Artistotle's sphere (l), Newton's oblate Spheroid (m) and Cassini's prolate spheroid (r).
Data: *Made with Natural Earth. Free vector and raster map data @ naturalearthdata.com.*
Source: Elisabeth Nelson

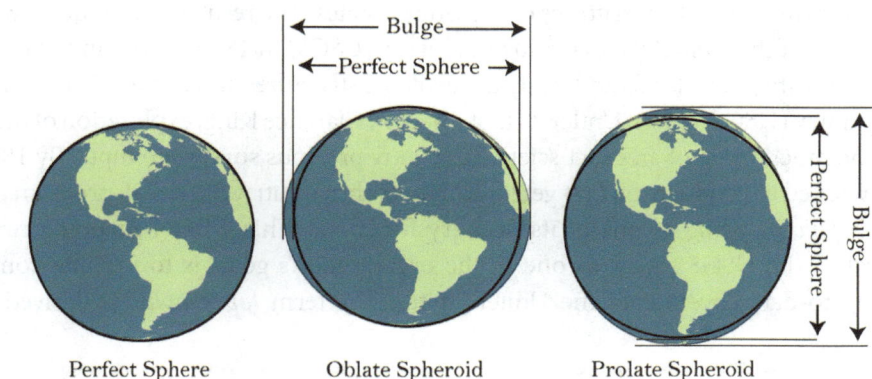

Such results presented a dilemma. First, even though the differences were small, they couldn't be easily dismissed. Isaac Newton's theory of universal gravitation also proposed that the Earth, due to centrifugal forces from rotating on its axis, was not spherical. Interestingly enough, however, Newton believed the Earth's shape was more like an *oblate spheroid* – flatter at the poles and slightly bulging at the equator (an egg on its side, if you will). Secondly, if Cassini were correct, and the lengths did differ, then the whole area would have to be re-surveyed and surveying methods refined to produce a truly accurate map.

To put this controversy to rest, the Académie sponsored two expeditions to widely separated latitudinal regions to find a definitive answer. They sent one team of surveyors to equatorial South America and another to the Arctic Circle. It took ten years, but in the end, both teams confirmed Newton's hypothesis: the Earth was not spherical after all, but really an oblate spheroid, with the poles slightly flattened in comparison to the equator. The difference is actually small (Figure 5.2 is exaggerated) and can usually be ignored when mapping at small scales; for example, one earth model widely used in mapping (the World Geodetic System) is a spheroid with only 21 km difference between the radius at the equator (6,378.137 km) and the radius at the poles (6,356.752 km). Of course, 21 km makes a big difference when the map you are making is a larger-scaled map to be used by others to calculate distances and directions.

With this new knowledge, the Académie had the Paris meridian resurveyed. Although Jean-Dominique Cassini passed away in 1712, before the entire country could be surveyed, his son, Jacques Cassini, continued the project, and in the late 1700s his grandson, Cesar-Francois Cassini de Thury, completed it. Known as the *Carte de Cassini*, the methods by which it was created became the standard in surveying techniques, and other countries later adopted them in the creation of their own surveys (Thrower, 1996).

Topographic Mapping and the United States Geological Survey

Between roughly 1500 and 1800, thousands of maps of the United States were produced; they ranged from crude sketches to surveys of limited scope (Brown, 1977). It wasn't until 1878, however, that Congress requested the *National Academy of Sciences* to devise a plan for surveying the territories of the United States. The result of this request was the formation of the *United States Geological Survey (USGS)* in 1879, whose initial purpose was to classify public lands and study the geologic structures and economic resources of the country (USGS, 2000). Under its first director, Clarence King, exploration of the west became much more of an exact science than any previous survey attempts. By 1882, as they worked to produce the first geologic map of the country, the USGS program added *topographic mapping* as one of its primary functions. This type of mapping remains central to the USGS today, as one of the organization's goals is to provide complete and up-to-date coverage of the United States. The term *topographic* is derived from

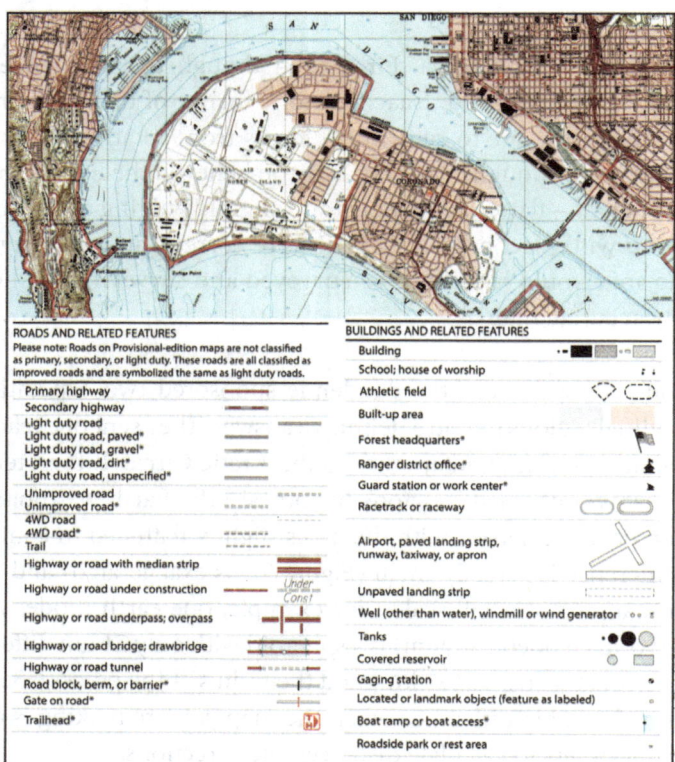

FIGURE 5.3
Example of USGS topographic map. *U.S. Geological Survey, Department of the Interior.*

the Greek *topographos*, which means "to describe a place". USGS topographic maps do just that (Figure 5.3); they describe prominent natural and cultural features using a cartographic symbol set on a backdrop of one or more *planar coordinate systems*, which allow you to locate the positions of represented features on the Earth's surface.

DATUMS AND THE GEOGRAPHIC COORDINATE SYSTEM

A planar coordinate system requires a three-dimensional surface that approximates the size and shape of our planet as its foundation. As we have seen, the Earth's surface more closely approximates an oblate spheroid than a true sphere. Mathematically, an oblate spheroid is an ellipse that is rotated on its shorter, semi-minor axis – the polar axis in Earth's case (Figure 5.4). The resulting three-dimensional surface often goes by the term *reference ellipsoid*. Its final shape is determined by 1) the length of the equatorial (or semi-major) axis **a** and 2) the length of the polar (semi-minor) axis **b**. Geodesists also specify reference ellipsoids by their inverse *flattening* (1/f), where **f = ((a-b)/a),** the proportion of difference between equatorial and polar axes relative to the equatorial axis length. When the difference in axis lengths is small, say 21 km, as is the case with Earth, the flattening that occurs is also very small (1/300), although still significant for large-scale mapping.

FIGURE 5.4
Parameters for specifying a reference ellipsoid.
Source: Elisabeth Nelson

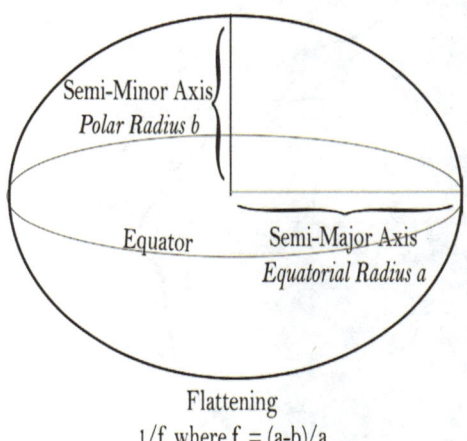

By the time the USGS began their topographic program, survey measurements had been recorded in several locations, with the length of a degree of latitude varying slightly from place to place. This is interesting, because it implies that a standard ellipsoid tied to a specific point on the Earth's surface varies in fit as you map different locations; depending on the local area being mapped, the ellipsoid's axis lengths and their proportion to one another must be modified to create a best-fit and to determine the most accurate coordinates. What might cause this? As it turns out, when surveyors were pinpointing these locations, the traditional equipment they used was oriented using the pull of gravity. This pull varies from place to place due to the Earth's rotation and the variation in composition and density of its crust – among other things. In other words, the Earth is not really an ellipsoid, either. Its true nature is rather lumpy, and goes by the name *geoid* (Figure 5.5).

FIGURE 5.5
A depiction of the Earth's geoid using gravity field anomalies.
NASA's Earth Observatory.

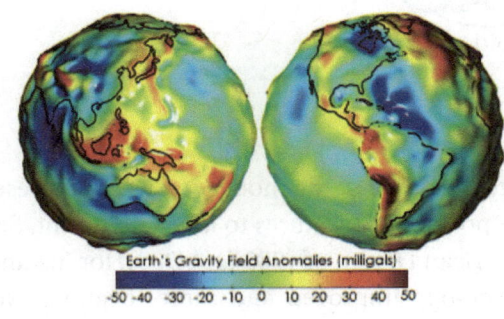

The geoid, however, is mathematically complex and cumbersome to use as a model for most surveys. Instead, different countries chose to use different 'best-fit' ellipsoids when conducting local surveys. In North America, for example, surveyors chose the *Clarke 1866 Ellipsoid*. They established a ground zero, or starting point for measuring locations of features, by tying the ellipsoid to the Earth's surface at Meades Ranch, Kansas. Using latitude, longitude, and the process of triangulation, they were then able to work outward from that location and measure a set of locations called *geodetic control points*. These points were located using benchmarks; they form a *geodetic datum* from which all other features can be located. This datum (Figure 5.6), known as the *North American Datum of 1927 (NAD27)*, consists of the ellipsoid, the origin of the coordinate system, and some 26,000 surveyed points in the United States and Canada (Bolstad, 2005). Its purpose is to describe locational, directional, and scale relationships in relation to Meades Ranch. Most large-scale maps of areas in the U.S. produced from the late 1800s through the late 1900s use this datum as the foundation for locating features within the country.

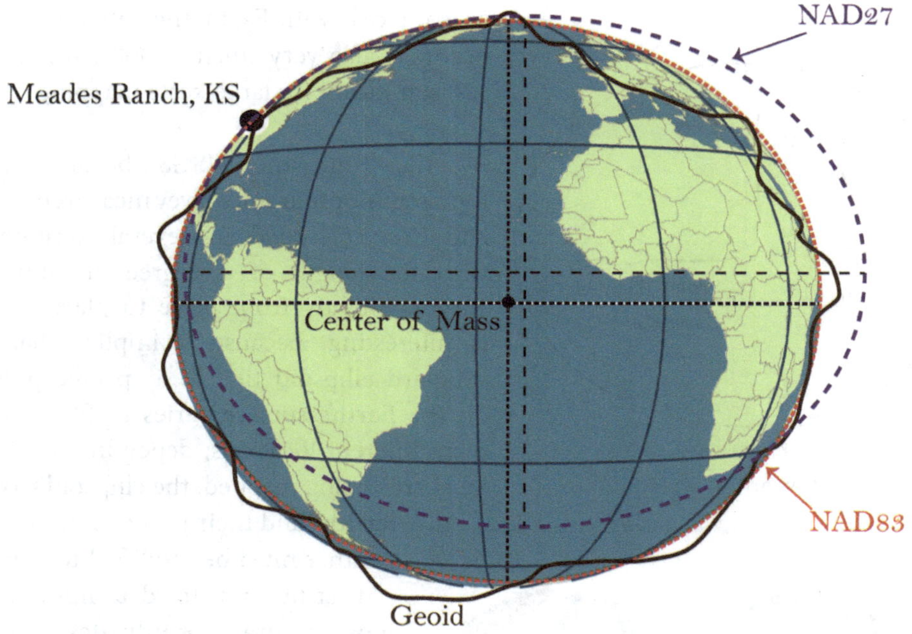

FIGURE 5.6
North American Datum of 1927, showing how the Clarke 1886 reference ellipsoid was tied to Meades Ranch, Kansas to create a 'best-fit' scenario for U.S. topographic mapping. Data: *Made with Natural Earth. Free vector and raster map data @ naturalearthdata.com.*

As you might suspect, when you change the shape of the model you use to represent the Earth, you also change the absolute positions of features to a certain extent. So it is with different datums. The North American Datum of 1983 (NAD83), for instance, was the successor to NAD27; as technology improved and more locations were surveyed, the National Coast and Geodetic Survey was assigned the task of updating our national maps. Instead of using Clarke 1866 as the ellipsoidal reference, they chose the *Geodetic Reference System 1980 (GRS80)* ellipsoid because technology had improved to the point that the local, fixed station approach of NAD27 was outdated. With NAD27, the ellipsoidal surface was tied to the Earth at one point, Meades Ranch, Kansas; thus, it was of limited use to anyone outside of North America. NAD83 ties the GRS80 ellipsoid to the center of Earth's mass, making it a datum that is global in use, as opposed to local (Figure 5.6). The result? Some locations in North America shifted by up to 1,312.3 feet, requiring the publication of new latitude and longitude values for some 250,000 control points. Of course, the features themselves did not really change location; rather, by moving to an updated datum, based on more accurate surveying technology, their locations in the new system are more precisely defined. This change is shown on older topographic maps by a dashed plus sign in the upper and lower left boundaries of the quadrangle, which indicates the map corner's position for the older NAD27 coordinates (Figure 5.7).

FIGURE 5.7

NAD27 tic marks on an updated USGS topographic quadrangle. *U.S. Geological Survey, Department of the Interior.*

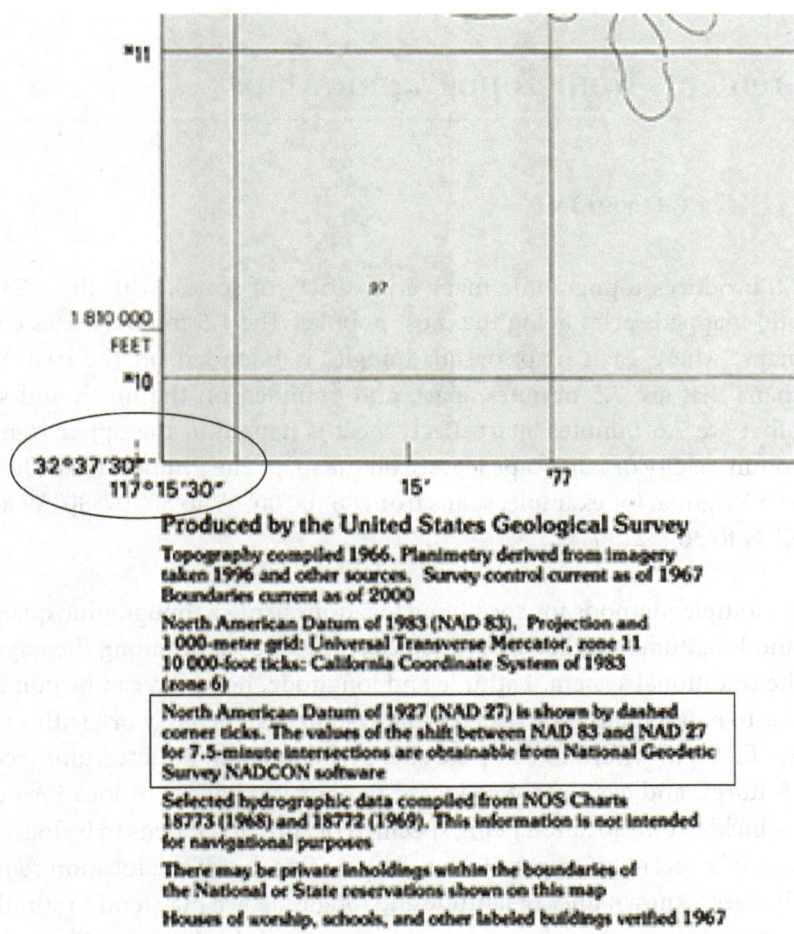

Produced by the United States Geological Survey

Topography compiled 1966. Planimetry derived from imagery taken 1996 and other sources. Survey control current as of 1967 Boundaries current as of 2000

North American Datum of 1983 (NAD 83). Projection and 1 000-meter grid: Universal Transverse Mercator, zone 11 10 000-foot ticks: California Coordinate System of 1983 (zone 6)

North American Datum of 1927 (NAD 27) is shown by dashed corner ticks. The values of the shift between NAD 83 and NAD 27 for 7.5-minute intersections are obtainable from National Geodetic Survey NADCON software

Selected hydrographic data compiled from NOS Charts 18773 (1968) and 18772 (1969). This information is not intended for navigational purposes

There may be private inholdings within the boundaries of the National or State reservations shown on this map

Houses of worship, schools, and other labeled buildings verified 1967

What does this mean for us, as users of these products and of other technological services? Take as an example a parcels layer created from a survey that used the NAD27 datum as its basis for recording the parcel locations (Figure 5.8). This layer has been overlaid on an aerial photograph of the same area, only the two layers do not align. The problem is that the aerial photograph is based on the NAD83 datum, so when the two are overlaid, you get a shift in feature positions, which at times, can be on the order of 200 meters (about 656 feet). Discrepancies such as this are quite common when compiling GIS projects, and the solution lies in using the correct software transformations to bring one layer into alignment with another by making sure all geographic coordinate systems match in the reference ellipsoid they use as their foundation for specifying latitude and longitude positions.

FIGURE 5.8
Misalignment of spatial data due to datum shifts. *Natural Resources Conservation Service, U.S. Department of Agriculture.*

Measurements from Topographic Maps

SPECIFYING LOCATION

The USGS produces topographic maps at a variety of scales, with the 1:24,000 (or 7.5 minute) mapped series being the most popular. The 7.5 minute series comprises 57,000 maps, where each map, or quadrangle, is bounded on the east and west by meridians that are 7.5 minutes apart, and bounded on the north and south by parallels that are 7.5 minutes apart. Each sheet is named in the upper right corner after a prominent city or landscape feature on the map. The Philpott Lake Quadrangle in southern Virginia, for example, spans from 80° 00'00" W to 80° 07' 30" W and from 36° 45' 00" N to 36° 52' 30" N.

There are multiple methods for specifying locations using a topographic quadrangle–latitude and longitude, marked by graticules and/or tic marks along the edges of the map, is the traditional system. Latitude and longitude, however, can be cumbersome for humans to manipulate. As a sexagesimal system designed to work with a spherical surface, we find it awkward to compute changes in degrees, minutes, and seconds for locating features, and even more awkward to convert changes in locations to linear distances quickly. If the location being specified or sought happens to be located at the intersection of a meridian or parallel, we may be fine, but if the location requires we estimate between known lines of latitude and longitude, we may tend to stumble. This is especially true when distances and directions are involved, as we will see shortly. Fortunately, there are other means of specifying locations and measuring distances and directions when using a topographic quadrangle. These systems are based on the idea of *Cartesian coordinates*, developed by a French philosopher/mathematician named René Descartes. Descartes used two perpendicular axes that cross at a central point as the basis of his plane. The central point is called the *origin*, and along with

FIGURE 5.9

Latitude and longitude
expressed as a planar
coordinate system.
Data: *Made with Natural
Earth. Free vector and
raster map data @
naturalearthdata.com.*
Source: Elisabeth Nelson

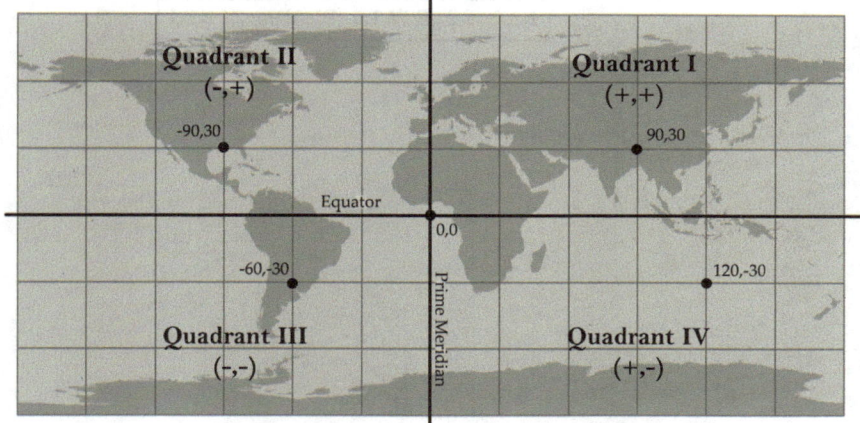

the two axes, it divides the plane into four quadrants, much like we saw when we transformed our geographic coordinate system to a plane using the Plate Carrée projection (Figure 5.9). Object positions can be determined east or west of the origin along the x-axis (the *easting coordinate*) and north or south of the origin along the y-axis (the *northing coordinate*). In GIScience, when working with navigational tasks, we are only interested in working in Quadrant I. This is because it is quicker and less error-prone to calculate positions, distances, and directions by working solely with positive numbers. How can we make this work, though, for any location on the planet? The answer lies in manipulating the coordinate system to create a customized system for navigational tasks.

State Plane Coordinate System. Developed in the 1930s, the State Plane Coordinate System (SPCS) is a customized grid system used at the state-level in the United States to record locations using Cartesian coordinates. The system was a boon for surveyors and engineers who could now complete their work without the need of spherical trigonometry. One key to the development of the system, however, was the need to minimize scale distortion; remember that as larger and larger areas are mapped, more distortion creeps in due to transforming our curved surface to a flat plane. To minimize the need to deal with curvature, it was decided to divide each state into one or more zones, and then create a separate coordinate system for each of these smaller regions (Figure 5.10).

Within each of these regions, positioning using this coordinate system is quite accurate, but outside each zone, accuracy declines, so as a coordinate system, the SPCS is not useful for national mapping applications. Each zone also has its own projection surface to minimize distortion even further: the Lambert Conformal Conic is used for zone shapes that have a predominant east-west extent like Tennessee, and the Transverse Mercator is used for zone shapes that have a predominant north-south extent, such as Georgia.

North Carolina's SPCS, for instance, is based on the Lambert Conformal Conic Projection (Figure 5.11). This projection has two standard parallels that help control distortion within the state. These parallels are specified so that they lay one-sixth of

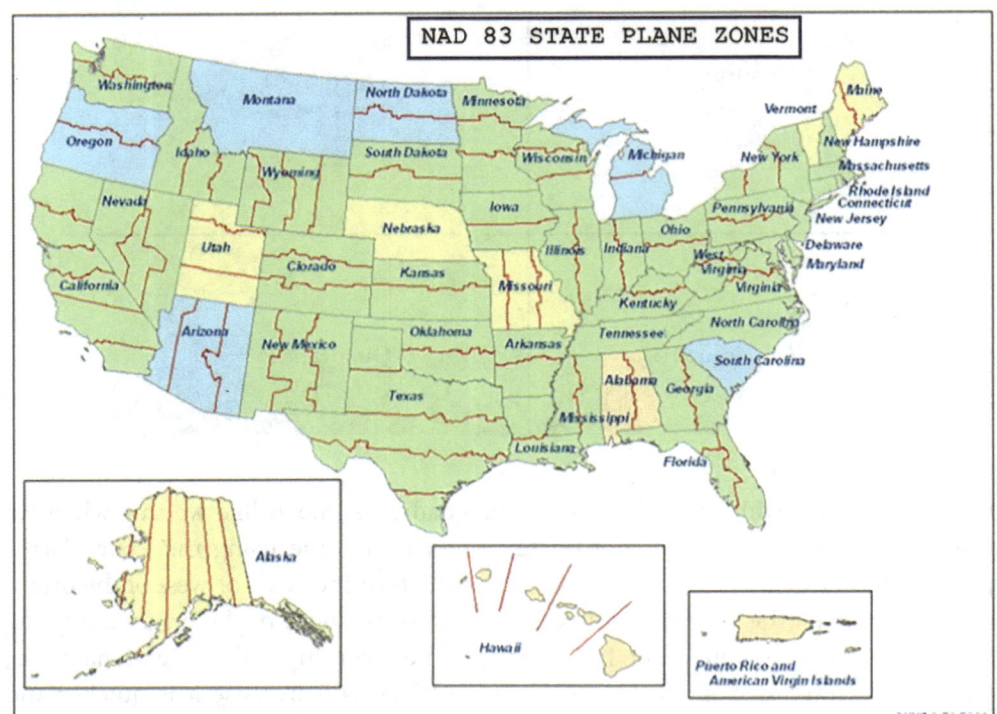

FIGURE 5.10
State Plane Coordinate System. *National Geodetic Survey, National Oceanic and Atmospheric Association.*

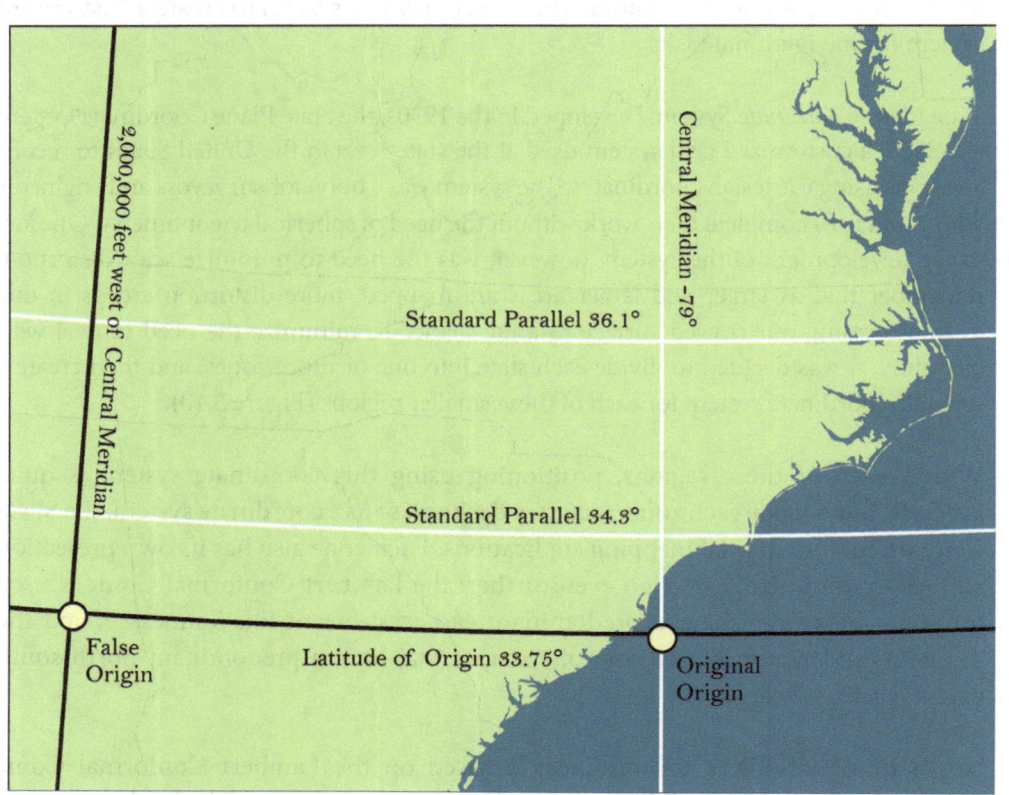

FIGURE 5.11
North Carolina's State Plane Coordinate System.
Data: *Made with Natural Earth. Free vector and raster map data @ naturalearthdata.com.*
Source: Elisabeth Nelson

EXPLORING GEOSPATIAL TECHNOLOGY

the zone width from the northern and southern zone boundaries of the state. A central meridian that is near the center of the zone is also specified. The central meridian serves as the starting point for a y-axis in this modified Cartesian system. Remember, however, that in GIScience, practitioners only want to deal with Quadrant I, where all coordinates are positive numbers. To make this work in the SPCS, GIScientists use a *false origin*, which is established west of the central meridian (2,000,000 feet for zones using the Lambert Conformal conic and 500,000 feet for those using the Transverse Mercator projection). The modified starting point for the x-axis is just south of the zone boundary, and is called the *latitude of origin*.

False origins, then, are created by adding numbers to the original origin of the projection. In doing so, the new origin is repositioned to the west and south of the original so that all coordinates within the mapped area become positive coordinates. To specify the location of a feature using the SPCS, you measure the distance in feet east and north of the false origin, and specify the location using that *easting* and *northing* – which act like an (x,y) coordinate—also noting the state and zone for the coordinate. As the x-coordinate, the easting gives the distance east from the false origin, while the northing (or y-coordinate) gives the distance north of it. Since you are only working with mapped features within that zone, recalibrating the coordinates to account for the false origin doesn't affect measurements, except to make them easier to calculate. On topographic quadrangles, grid lines for the SPCS are marked using black tics along the edges of the map.

If the feature you are locating lies on one of the intersections of these gridlines, the easting and northing coordinate can be specified with little extra work. If, however, the point of interest is located somewhere between gridlines, as the Lewis and Clark school is in Figure 5.12, then approximating the location requires using methods of interpolation. To locate the school, for example, you would:

1. Find the nearest gridlines that are south and west of the feature
   Northing and easting values are posted at one tic along each side of the topographic quadrangle. Additional unlabeled tic marks are posted for every 10,000 feet for the 1:24,000 series.
2. Using a strip of paper, measure the distance from each gridline to the point by making a tic mark on the strip at each point
3. Align the strip along the map's scale bar to determine the ground distance between each gridline and the feature
4. Add the gridline values to each measured distance to determine the easting and northing coordinate for your point

For Lewis and Clark school, this would result in a coordinate location of approximately: (1,154,752 ft. E, 924,541 ft. N, OR), with the final accuracy influenced by measuring accuracy.

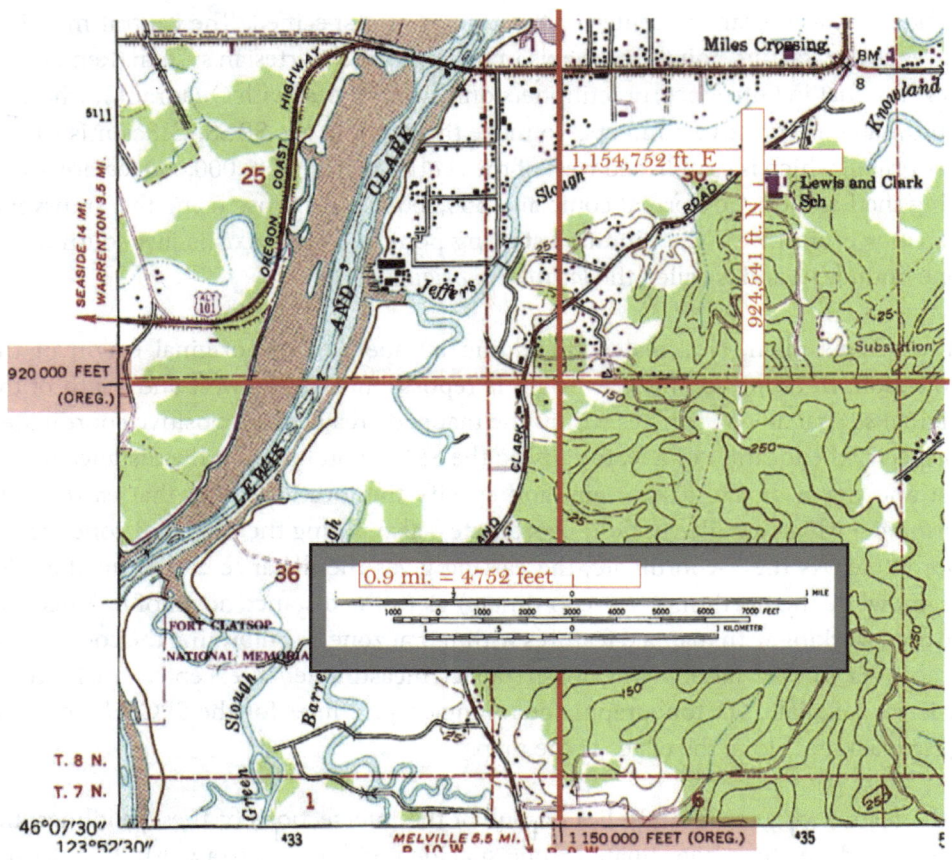

FIGURE 5.12
Locating a feature using
the SPCS. *U.S. Geologi-
cal Survey, Department
of the Interior.*

Universal Transverse Mercator System (UTM). While the SPCS was defined only for the U.S., with more modern technology we can now create a worldwide grid coordinate system. The *Universal Transverse Mercator (UTM) system* is one such system (Figure 5.13). It works by dividing the Earth into 60 north-south zones, each of which extends from 84° north to 80° south. Each zone, numbered 1-60 beginning at the International Date Line and moving east, is based on a transverse Mercator projection with its own specified central meridian that acts as a standard line. Because each zone is only 6 degrees wide, the projected data have a high degree of accuracy. The polar regions are covered by their own complementary grid coordinate system, called the Universal Polar Stereographic System. Both systems measure distances in meters north and east of a zone origin.

Each UTM zone has a separate false origin for its Northern and Southern Hemisphere components. The false easting for these origins is defined similarly – it is always 500,000 meters west of the central meridian for the zone in question. Thus, the central meridian in each of the 60 zones is marked 500,000 meters, and locations west of that meridian range from 0–500,000 meters. Those locations east of the central meridian have eastings greater than 500,000 meters. The false northing is hemisphere-

FIGURE 5.13

Universal Transverse Mercator Coordinate System.
U.S. Geological Survey, Department of the Interior.

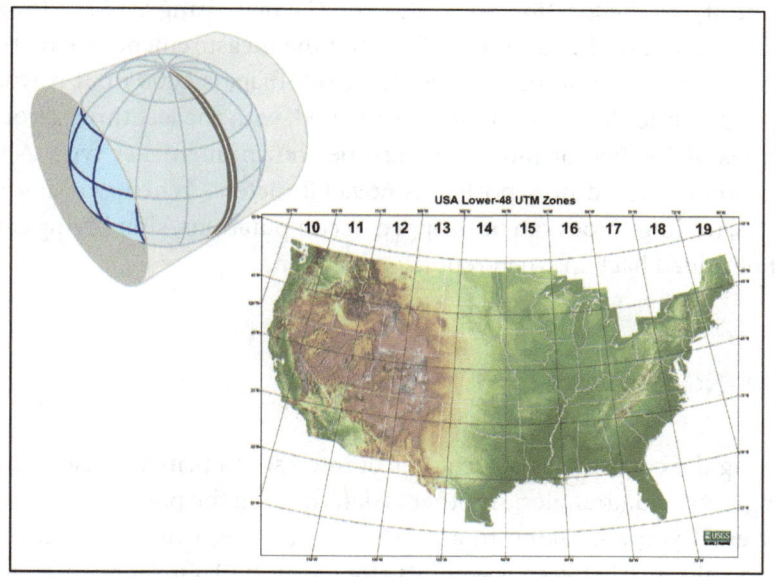

dependent. In the Northern Hemisphere, the northing origin (0 meters) is the equator, so that all northing coordinates are greater than 0 meters. A location is specified by giving the false easting coordinate, the false northing coordinate, the zone, and the hemisphere (Figure 5.14). When working in the Southern Hemisphere, the equator is assigned the value of 10,000,000 m, displacing the false origin to the southern polar region, so that all coordinates, again, may be reported as positive numbers.

FIGURE 5.14

Feature location using the UTM coordinate system.
Made with Natural Earth. Free vector and raster map data @ naturalearthdata.com.
Source: Elisabeth Nelson

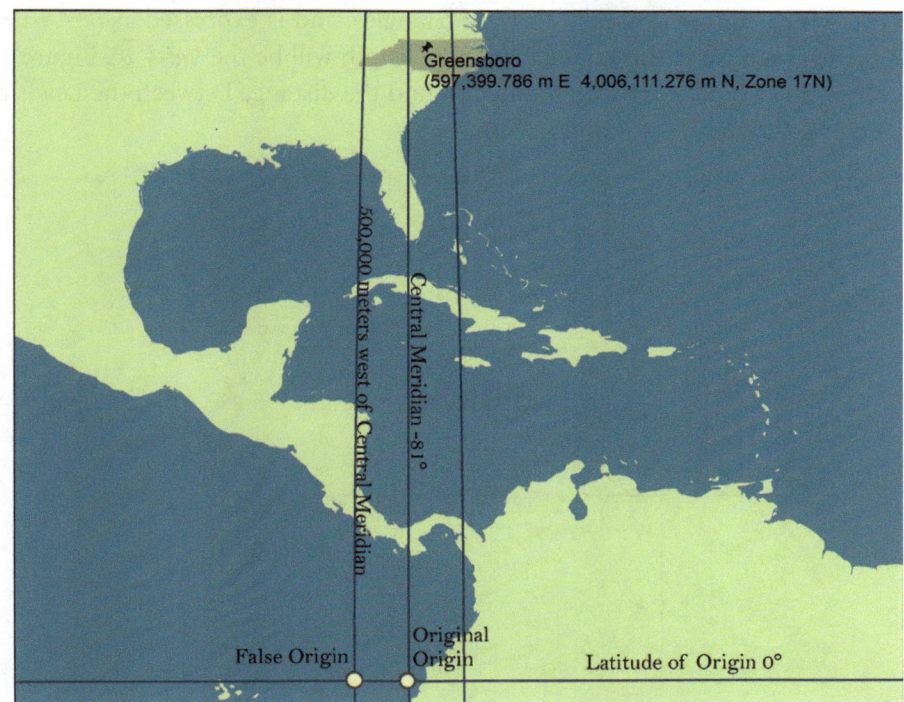

Each zone can then be mapped using multiple map sheets, depending on the scale needed. Specifying the location of a feature on the map using UTM follows the same procedures as those outlined for the SPCS, but the measurements are metric and the grid interval, using a 7.5 minute series (1:24,000) map, is 1000 m. It is worth noting that most UTM grid designations are abbreviated, with the last three zeroes omitted. Designations of 100,000 or more may also be shown in smaller type. A reading of [6]69, for example, would be reported as 669,000 meters. When working with these coordinates, trailing zeroes can be dropped in computations, simplifying calculations, but must be added back in when reporting locations.

ESTIMATING DISTANCES

Determining the coordinates for features of interest is a primary task when working with topographic quadrangles. Another is determining the physical distance between two features. If you are hiking in an area, for example, you might be interested in knowing the distance between waypoints along your trail. On large-scale maps, where the Earth's curvature is accounted for using conformal map projections and ellipsoidal datums, there are several ways to estimate these distances. In Chapter 2, we examined the representative fraction, the verbal statement, and the bar scale as different ways of accomplishing this task.

Another, perhaps more precise way to do this, would be to use the *Distance Theorem* in conjunction with the planar coordinates of the two features of interest. This is the process Global Positioning Systems (GPS, which is discussed later) and GIS use to calculate distances. The Distance Theorem is a derivation of the familiar *Pythagorean Theorem*, which concerns itself with right triangles, and is expressed as $a^2 + b^2 = c^2$. If c is the hypotenuse of this triangle, then its length will be the $\sqrt{a^2 + b^2}$. Figure 5.15 shows how this would work if we wanted to find the distance between the Lewis and Clark School and the Navy Radio Station.

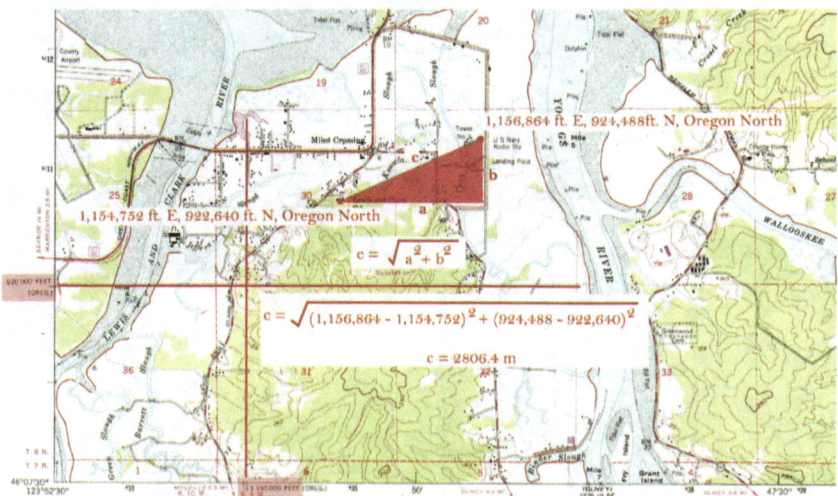

FIGURE 5.15
Determining the distance between two features using the UTM coordinate system.
U.S. Geological Survey, Department of the Interior.
Source: Elisabeth Nelson

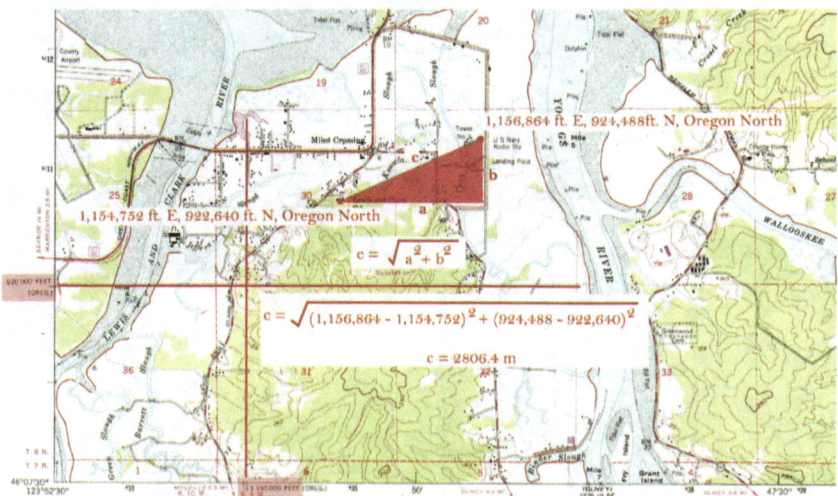

If we make the distance between these two locations the hypotenuse (c), we can create a right triangle by connecting the easting coordinates of the two locations and the northing coordinates of the two locations ($x_2 - x_1$). The difference between the easting coordinates of these two locations is one known leg of the triangle (a); the difference between the northing coordinates of the two locations ($y_2 - y_1$) gives you the length of the second leg of the triangle (b). When we substitute these distances in the original Pythagorean Theorem, the Distance Theorem results: ($c = \sqrt{(x_2 - x_1)^2 + (y_2 - y_1)^2}$). You can use any planar coordinate system, and you should get very similar results.

While this is a relatively straightforward process, it should only be used when your locations are both within the same zone. Once you involve more than one zone in your calculations, you increase your measurement errors, because the size of the region (and its associated curvature) grows beyond what can be handled accurately with planar coordinates.

DETERMINING DIRECTION

Geographic North. Most map users working with distances between locations will also want to know the direction they need to travel from their starting location to their destination. To determine that direction, a frame of reference is needed. Geographically, that reference is typically a north-south baseline that is oriented to true, or geographical, north. A *true north* reference line points to the north pole of Earth's axis of rotation. A meridian on a globe, then, from any point on the planet to the North Pole is a true north reference line.

North is one of the four cardinal directions used to orient yourself geographically on Earth. Together, north, south, east, and west are called *cardinal directions*, and are possibly the primary type of direction with which you are familiar. The geometric foundation (our north-south baseline) of this system, however, allows us to specify direction more precisely as an angular deviation from that reference line. As you face north, you can determine the direction to any other point by extending an imaginary line from you to that point, then measuring the angle made by the intersection of that direction line with the baseline. This form of direction is known as *azimuth* and it is measured in angular degrees clockwise from *true north* (Figure 5.16). Thus, as you move clockwise from true north, east is at a right angle and is designated as 90°, south is 180°, and west is designated as 270°.

On maps, true north is either assumed to point to the top of the map, when the map is being held in such a position that the names are right-side up, or it is indicated by a north arrow. On a topographic quadrangle, it is indicated by a north arrow marked with a star (Figure 5.16). If you are trying to find true north while navigating the environment, you have several options, from using astronomical observations to observing shadows or the sun's location in the sky. By far the easiest in today's world, however, is to use a system of satellites popularly known as *GPS* or the *Global Positioning System*. This network of satellites was developed by the U.S. Department

FIGURE 5.16
Determining the azimuth
between two features.
*U.S. Geological Survey,
Department of the Interior.*
Source: Elisabeth Nelson

of Defense and was deployed in the 1990s. Containing some 32 satellites orbiting the Earth, each satellite transmits a radio signal to Earth that can be read by a GPS receiver to calculate not only your position on the planet, but also distances and directions to distant locations. Once you have established your location, calculating distances and directions to another feature requires you only to input the location of that feature – the GPS does the remaining mathematical work.

Grid North. Since our goal is to familiarize ourselves with how maps work, we will focus here on determining directions using topographic maps. True north is actually just one of three north-south baselines you will find on a topographic quadrangle (Figure 5.16). More commonly used in map work are both *grid north* (the north arrow marked with GN) and *magnetic north* (the north arrow marked with MN). Grid north is used in conjunction with grid coordinates, such as UTM and SPCS. Planar coordinate systems are artificial, though, so working with grid north will be artificial as well, meaning that the directions measured will only be useful in the context of the map sheet with which you are working. Grid north is the direction of a grid line that is parallel to the central meridian of the map's projected coordinate system. Typically, this line won't align with true north because it ignores the fact that meridians, in reality, curve and converge at the poles. It is, however, a useful designation for measuring and computing directions from maps, and is the way that many GIS calculate azimuth readings.

Figure 5.17 shows an example of determining this type of direction from Moore's Knob at Hanging Rock State Park to Danbury, N.C. First, the grid coordinates of each feature are determined by following the same steps outlined earlier in this chapter. Once these are known, a right triangle can be formed by connecting the two locations with a straight line and using the easting and northing coordinates to determine the distances of the other two legs of the triangle. Right triangles, in addition to being associated with the Pythagorean Theorem, are also linked to several trigonometric functions useful for calculating angles from distance ratios.

FIGURE 5.17

Determining a grid
direction between two
features in Quadrant I.
*U.S. Geological Survey,
Department of the
Interior.*
Source: Elisabeth Nelson

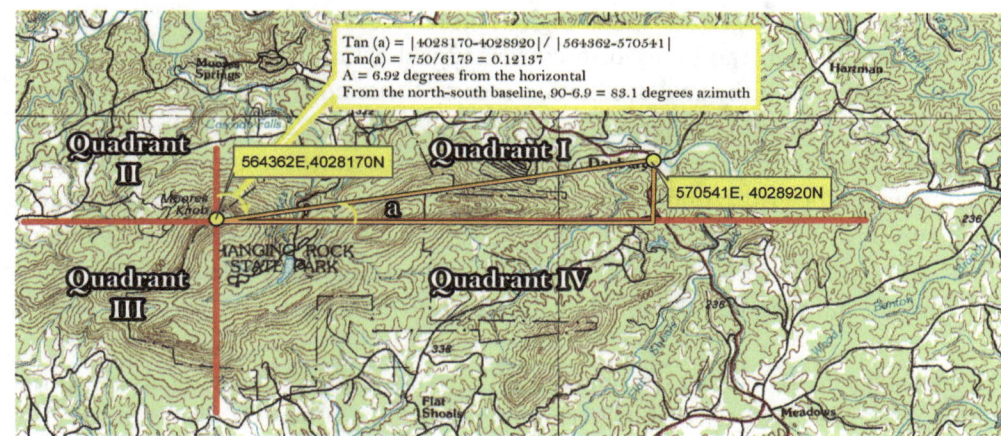

In this case, the *tangent* function is used to calculate the grid azimuth between Moore's Knob and Danbury. Measuring from a horizontal baseline, the tangent function computes an angle from the ratio of the length of the triangle's side opposite of the angle in question to the side of the triangle adjacent to the same angle. It is determined by using this formula:

$$\text{Tan}(a) = |y_1 - y_2| / |x_1 - x_2|$$

Once the ratio is determined, the associated angle (a) is calculated by taking the inverse tangent function. Since the grid azimuth should be measured from a line pointing north as the starting point instead of a horizontal baseline, however, we must make one final modification. To convert the angle to a grid azimuth, we must subtract it from 90° when working in Quadrant I, add 90° to the angle when working in Quadrant IV, subtract the angle from 270° for locations in Quadrant III and add 270° to the angle for those angles in Quadrant II.

If, instead, we were asked to determine the azimuth from Moore's Knob to Flat Shoals (Figure 5.18), which is in Quadrant IV, we would need to *add 90° to the initial angle* to account for moving from the north-south baseline through Quadrant I to Quadrant IV.

FIGURE 5.18

Determining a grid
direction between two
features in Quadrant IV.
*U.S. Geological Survey,
Department of the
Interior.*
Source: Elisabeth Nelson

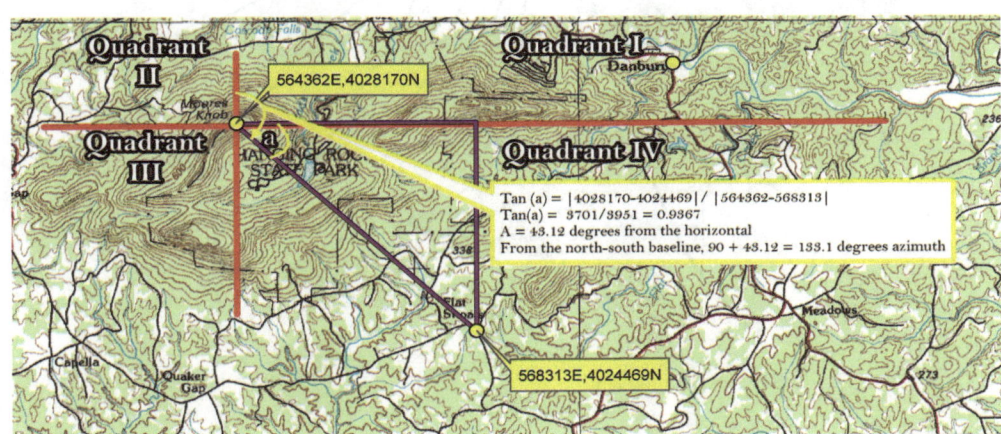

The angular difference between grid north and true north, as shown in the declination diagram on the topopgraphic quadrangle (Figure 5.16) is called the *grid declination*. The grid declination can be used to convert a grid azimuth to a true azimuth or a magnetic azimuth. If the grid declination is east of true north, for example, then you would subtract the grid declination from the grid azimuth to arrive at the true north azimuth. Likewise, if the grid declination is west of true north, you would add the grid declination to the grid azimuth to reach a true north azimuth. In Figure 5.17, the grid declination is 18' E of true north. If our grid azimuth from Moore's Knob to Danbury is 83°, then to convert that azimuth to a true north reading, we would need to subtract 18' from the number; 82° 42' would be the true north azimuth.

Magnetic North. True north and grid north are two of the primary forms of north-south reference lines used in determining directions. There is also a third that bears mentioning: *magnetic north*. Used in conjunction with magnetic compasses, magnetic north points to the earth's magnetic pole, which is aligned north and south, but does not align with true north. On top of not aligning with true north is the fact that the magnetic north pole actually shifts, or changes position through time. Currently, it is located in Canada, some 800 miles south of the true North Pole. The difference between the location of true north and magnetic north is called the *magnetic declination*, and is noted on USGS maps with a MN at the top of the north arrow in the declination diagram (Figure 5.16). Because the magnetic pole shifts over time, the date a map is produced is critical to relying on the magnetic declination information posted. There are maps designed to show magnetic declination over large areas, such as the U.S.; they display variations in angular difference between true and magnetic north using *isogonic lines*, or lines of equal declination angles (Figure 5.19). It is interesting to note that there is a line of no angular difference, the *agonic line*, and that it is currently in the east central part of the U.S. In other words, navigating with a magnetic compass in this part of the U.S. is relatively problem-free, compared to the northwest or northeast, where magnetic declination is more substantial, requiring navigators to compensate when trying to plot a course for true north.

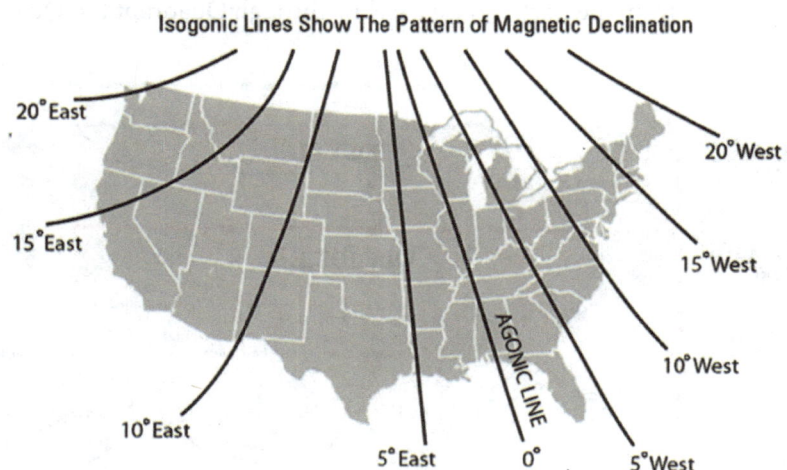

Isogonic Lines Show The Pattern of Magnetic Declination

FIGURE 5.19
Determining a grid direction between two features.
U.S. Geological Survey, Department of the Interior.

Symbology and Topographic Maps

Like reading a page of text, you must be familiar with the meanings of the "letters, words, and punctuation" of the cartographic alphabet to read and understand a map (Miller, et al., 1989). The map features that are typically noticed at first glance on a topographic quadrangle are areal, and symbolized by variation in *color hue*. Areas of vegetation, for example, are visualized using a green hue, whereas water features are associated with blues, and built-up environments with reds. In some instances, changes to the original topographic sheet, which were derived from interpreting aerial photographs, are visualized using a purple hue. There are also point features, such as buildings, campgrounds, schools, and churches; these are typically differentiated by changing the *shapes* of the symbols. Linear features, like areas, are categorized by color hue: roads are red, railroads and political boundaries are black, streams and rivers are blue.

The symbol that differentiates the topographic map from most others, however, is the *contour* (Figure 5.20). Brown in color, the contour is an imaginary line that represents a ". . . narrow band of places sharing roughly the same elevation above a datum, or fixed reference elevation, usually mean sea level" (Monmonier and Schnell, 1984:75). The contour adds a third, vertical dimension to the topographic map that is not present on the traditional *planimetric* map, which does not attempt to show elevational changes. Where the purpose of the planimetric map is to show where geographic features are and what their relationships are to one another, the topographic map not only also accomplishes this, but also provides symbology for visualizing 3D changes in terrain via *contour shapes* and *spacings*.

FIGURE 5.20
Example of contour lines on a topographic quadrangle. *U.S. Geological Survey, Department of the Interior.*

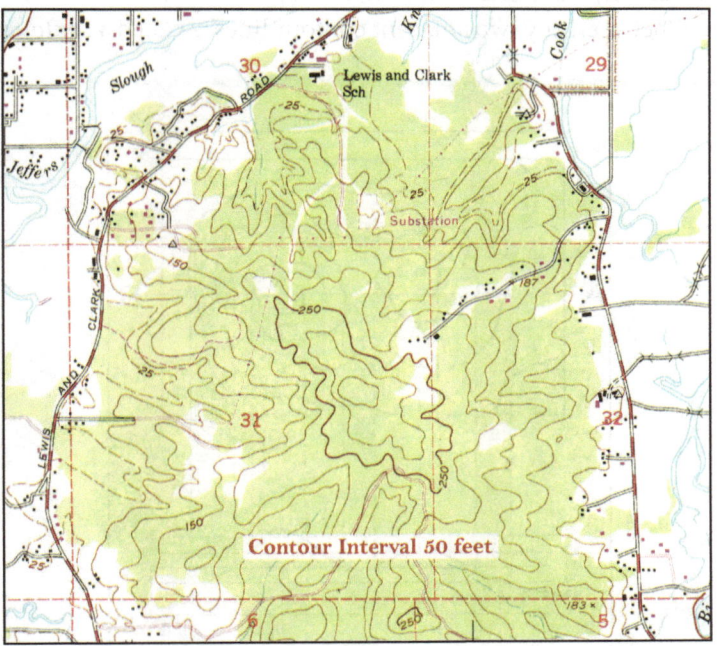

VISUALIZING TERRAIN

Vertical Datums. Similar to the horizontal datums used to pinpoint coordinate locations on the planet, there are also *vertical datums* that we use to measure elevation, or height above mean sea level. One way to think of mean sea level is as the average height of the ocean's surface when all high and low tides are taken into account. Many world maps use a mean sea level datum tied to the global geoid as their reference datum, but there are also other, local datums in use. For example, on some USGS topographic sheets you might see a reference to the *National Geodetic Vertical Datum of 1929 (NGVD29)* or to the *North American Vertical Datum of 1988 (NAVD88)*. Elevations made from these datums, which are based on local measures of mean sea level, will, naturally, vary not only between the two datums, but also from those mapped using a global geoid. GPS adds yet another option, as it calculates ellipsoidal heights above and below the WGS84 ellipsoid surface it uses. These measurements, however, should not be confused with elevations because they are not measured relative to mean sea level. To use these measurements as elevation values requires a conversion process. Like the choice of horizontal datum, then, the choice of vertical datum used in mapping is also important in defining what level of accuracy you can expect to obtain from your maps and mapping projects.

The Contour Line. Contours are used to mentally visualize the three-dimensional terrain that has been mapped, but the symbology and how to use it for that purpose may not be readily apparent to someone unfamiliar with it. Contours are probably best understood when connected to an actual three-dimensional model of the surface that was mapped (Figure 5.21). Theoretically, a contour line is simply the trace outline of a plane, parallel to the land surface, that passes through that surface. This trace represents all those locations that have approximately the same elevation. Since any one location can only have one elevation, contour lines will never cross one another. The increment between any two adjacent contour lines is called a *contour interval*, and

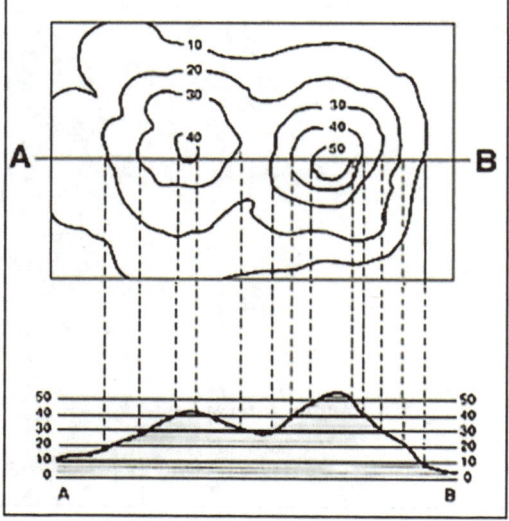

FIGURE 5.21
Concept behind contour lines. *U.S. Geological Survey, Department of the Interior.*

it represents the vertical change in elevation between the lines. For the purposes of the mapping technique, the change in elevation is assumed to be linear and uniform from one contour to the next. The contour interval is stated on USGS topographic maps, but you can also determine the interval on your own by locating two *index contours*, which are labeled, and subtracting the difference in the elevations, then dividing that difference by the number of "bands" created by the intermediate contours running between them (number of intermediate contours + 1).

This idea is easily seen in Figure 5.22. The index contours of 500 and 650 feet have 2 intermediate contours (3 bands) between them. Subtracting 500 from 650 leaves 150; 150 divided by 3 gives the map a contour interval of 50 feet. Index contours typically occur every fifth contour; they are the only labeled contours, a practice aimed at reducing map clutter. These labeled contours are also drawn using a slightly wider line so that it stands out relative to the other contours. Examples of other contour designs are shown in Figure 5.23. All provide additional information about the landscape, although some may be more commonly seen than others. *Supplementary contours*, for example, are used in areas where there are only slight changes in elevation to provide the map user with more detail in that particular area of the map. On USGS topographic maps, these lines are similar to intermediate or unlabeled contours, only lighter in value. Also fairly common are *depression contours*, which have tic marks added to the contours. These tics are drawn at right angles to the contour and point downhill, indicating that the area is a depression instead of a hill. Note that the outer depression contour (the first one marking the depression) will have the same elevation as its adjacent intermediate or index contour (Kimerling, et al., 2009).

FIGURE 5.22
Determining the contour interval. *U.S. Geological Survey, Department of the Interior.*

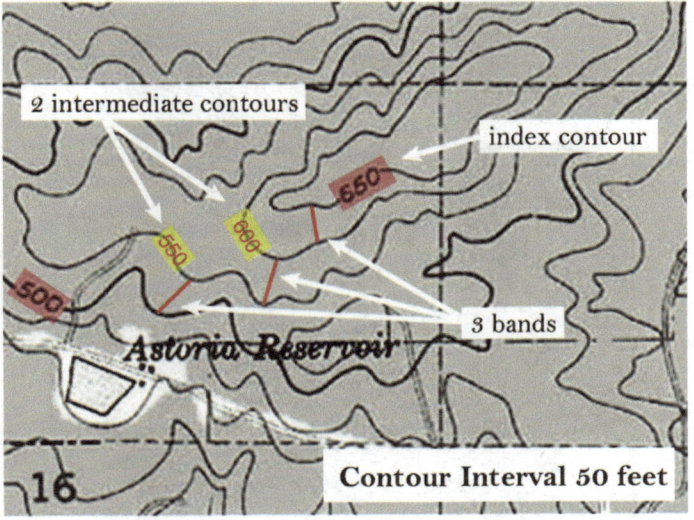

FIGURE 5.23
Types of contour lines.
*U.S. Geological Survey,
Department of the
Interior.*

CONTOURS

Topographic

Index ———6000———

Approximate or indefinite

Intermediate

Approximate or indefinite

Supplementary

Depression

Interpreting Contours. The shapes of contours and their spacings relative to one another are keys to visualizing the changing terrain (Figure 5.24). Contours that form closed loops, for example, indicate either a summit (hill) or a depression, depending on the type of contour line marking the area. The smaller loops representing a summit, then, will have the higher elevations; the larger loops are lower in elevation and are, thus, downhill. Depressions, on the other hand, have smaller loops that are lower in elevation and downhill from the larger loops.

If you see contours that form v's, then you are in a valley with a stream. The v shows where the contour crosses the stream, and the tip of the v points upstream towards the headwaters or uphill. The open end of the v points downstream or downhill. V's can occur whether or not the stream is actually mapped. Saddles are low points, or dips, in a ridge, which is a linear range of high elevations with little variation.

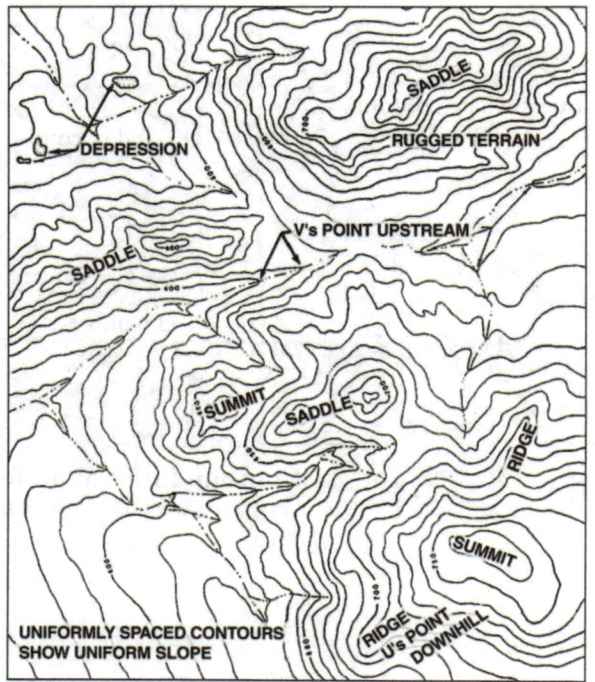

FIGURE 5.24
Interpreting contour
shapes and spaces.
*National Wildfire
Coordinating Group.*

The spacing of contours in an area also provides information. Closely spaced contours, for example, are indicative of steep hills or rugged terrain; change in elevation is occurring rapidly over a small horizontal distance. Widely spaced contours, on the other hand, still have the same contour interval, so change in elevation is occurring less rapidly over a longer horizontal distance. If the contours are uniformly spaced, then the change in slope over that area will also be uniform. Slope is the term for the change in elevation that occurs over a given horizontal distance. You can infer slope, as we have been discussing, by visually interpreting contour spacings; you can also estimate it numerically between two given points. To estimate it, use the following five steps (National Wildfire Coordinating Group, 2007):

1. Determine the contour interval of the map
2. Determine the elevation of the two points in question
3. Subtract the lower elevation from the higher to determine the *rise* or vertical distance between the points
4. Measure the *run* (horizontal) map distance between the two points
5. Use the formula ($\frac{Rise}{Run}$ x 100) to determine the slope percentage.

Additional Readings

Alder, K. 2002. *The Measure of All Things: The Seven-Year Odyssey and Hidden Error That Transformed the World.* Free Press: New York, New York.

Bogens, M., Durfee, M., Gardner, L., and R. Streeper. 2007. *Basic Land Navigation.* National Wildfire Coordinating Group: Boise, Idaho.

Bolstad, P. 2005. *GIS Fundamentals.* 2nd ed. Eider Press: White Bear Lake, Minnesota.

Brown, L.A. 1977. *The Story of Maps.* Dover Publications, Inc: New York, New York.

Campbell, J.C. 1984. *Introductory Cartography.* Prentice-Hall, Inc.: Englewood Cliffs, New Jersey.

Dent, B.D., Torguson, J.S., and T.W. Hodler. 2009. *Cartography: Thematic Map Design.* 6th ed. McGraw-Hill: New York, New York.

Faul, S. *Triangulation.* Retrieved from http://www.mundi.net/locus/locus_008/, May 2011.

Harvey, F. 2008. *A Primer of GIS: Fundamental Geographic and Cartographic Concepts.* The Guilford Press: New York, New York.

Kimerling, A.J., Buckley, A.R., Muehrcke, P.C., and J.O. Muehrcke. 2009. *Map Use: Reading and Analysis.* 6th ed. ESRI Press: Redlands, California.

Miller, V.C., and M.E. Westerback. 1989. *Interpretation of Topographic Maps.* Merrill Publishing Co.: Columbus, Ohio.

National Academy of Sciences. 2010. Founding of the National Academy of Sciences. In *The National Academies.* Retrieved August 2010 from http://www7.nationalacademies.org/archives/nasfounding.html.

National Wildfire Coordinating Group. 2007. *Basic Land Navigation.* NWGC: Boise, Idaho.

O'Connor, J.J., and E.F. Robertson. 2008. Jean Picard. In *The MacTutor History of Mathematics Archive.* Retrieved August 2010, from http://www-history.mcs.st-andrews.ac.uk/Biographies/Picard_Jean.htm.

Robinson, A.H., Morrison, J.L., Muehrcke, P.C., Kimerling, A.J., and S.C. Guptill. 1995. *Elements of Cartography.* 6th ed. John Wiley & Sons, Inc: New York, New York. Sickle, J.V. 2004. *Basic GIS Coordinates.* CRC Press: Boca Rotan, Florida.

Shodor. 2010. Cartesian Coordinate System. Retrieved August 2010 from http://www.shodor.org/interactivate/lessons/CartesianCoordinate.

Slocum, T.A., McMaster, R.B., Kessler, F.C., and H.H. Howard. 2005. *Thematic Cartography and Geographic Visualization.* 2nd ed. Pearson Education, Inc: Upper Saddle River, New Jersey.

Snyder, J. 1993. *Flattening the Earth.* University of Chicago Press: Chicago, Illinois.

Taylor, L.H. *Navigation: Near Shore Mapping- Adding Underwater Objects To the Contour.* Retrieved from http://www-personal.umich.edu/~lpt/mapping2.htm, May 2011.

Thrower, N.J.N. 1996. *Maps and Civilization: Cartography in Culture and Society.* University of Chicago Press: Chicago, Illinois.

USGS. 2000. Geological Surveys Before the Civil War. In *US Geological Survey Circular 1050.* Retrieved August, 2010 from http://pubs.usgs.gov/circ/c1050/before.htm.

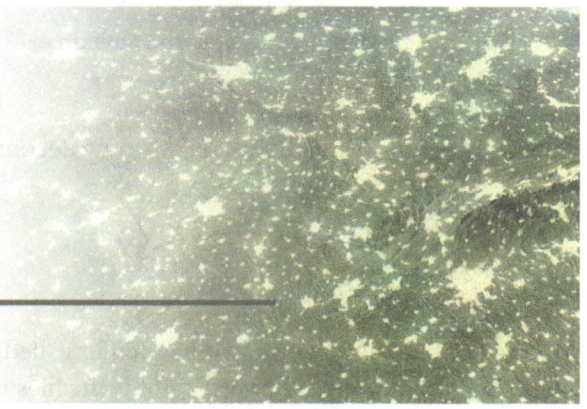

Exercise 5
Digital Topographic Mapping

In Exercise 5, you will use topographic maps and elevation data to work with planar coordinate systems and explore the visualization of elevation.

OBJECTIVES

- Identify map collar information on a topographic quadrangle
- Specify feature locations using 2D coordinate systems
- Estimate distances between geographic features
- Interpret elevation changes between contour lines

SOFTWARE INFORMATION

Introducing ArcMap uses Esri's **ArcGIS 10.6** software. **ArcGIS** is a commercial GIS package, available in geospatial labs affiliated with schools that have a campus-wide site license for the software. Instructors at these campuses may also request 1-year student versions of the software at http://www.esri.com/landing-pages/education-promo.

DATA

The data for the exercises is available from the Kendall Hunt Student Ancillary site. See the inside front cover for access information. You may also be directed to download the data from a different location by your instructor.

1. Download the data as instructed by your instructor
2. Save the file to a location where you have read/write privileges (USB key, home computer, class server)

The file you just saved is in a compressed (*.zip) format, and was created using **7-zip** freeware (http://www.7- zip.org). To use the data for this exercise, you must decompress it using the same software.

3. Locate the zipped file that you saved
4. Right-click the file
5. Choose *7-Zip - Extract Files*
6. Click *OK* to create a folder with the decompressed data

Open the Map Document

Your extracted files are grouped in a folder called **Chapter05**. This folder is your workspace folder and your map's **Home Folder** in ArcMap. It holds the files needed to display and manage the geographic information in ArcGIS for Exercise 5.

1. Start **ArcMap**. You can access ArcMap using several different methods, depending on your environment. You can start by clicking the Windows start menu icon at the bottom left corner of your screen, then navigate to the ArcMap icon in the ArcGIS folder, or search for ArcMap in the search window by typing "arcmap".
2. In the *ArcMap - Getting Started* dialog box, click *Browse for more* under *Existing Maps* (Figure 5.1)

FIGURE 5.1

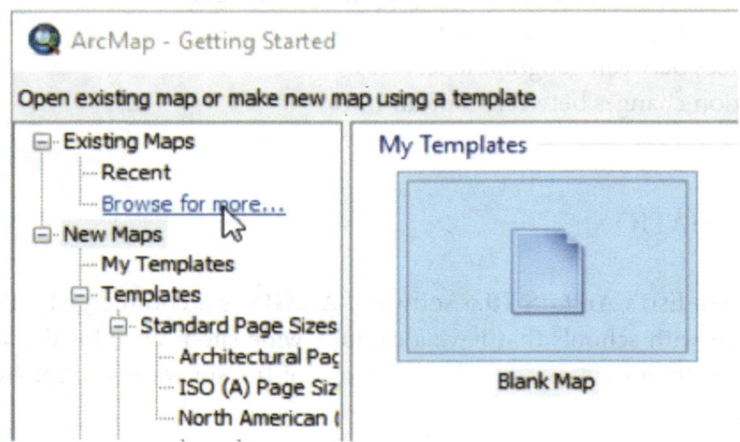

3. Navigate to the location where you saved the *Chapter01* data folder
4. In the folder, double-click **Exercise_05.mxd** to open the map document. **ArcMap** should open with a map that looks like Figure 5.2

FIGURE 5.2
Source U.S. Geological
Survey, Department of
the Interior

Identify Map Collar Information

USGS topographic quadrangles are composed of the map and the map collar. The map collar is the border that surrounds the map; it includes supplementary information crucial to working with the map.

Lines of latitude and longitude bound the sides of the USGS quadrangle. Answer the following questions about these boundaries:

QUESTION 5.1

The latitudinal line that forms the northern boundary of this quadrangle is 36° 52' 30" N.

  a. True
  b. False

QUESTION 5.2

The latitudinal line that forms the eastern boundary of the quadrangle is 109° 00' 00" W.

a. True
b. False

QUESTION 5.3

This quadrangle is a part of which map series?

a. 15 minute topographic series
b. 1:100,000 scale series
c. 7.5 minute topographic series
d. 1:250,000 scale series

QUESTION 5.4

The quadrangle located directly north of this quadrangle is:

a. Rocky Point
b. Cow Butte
c. Aneth SE
d. Pastora Peak

QUESTION 5.5

This quadrangle is part of which UTM zone?

a. 12
b. 14
c. 13
d. 11

QUESTION 5.6

The grid lines shown on this quadrangle are for the UTM grid system. What is the ground distance between adjacent grid lines in this system?

s.   10,000 meters
b.   100 feet
c.   1,000 meters
d.   100,000 feet

Coordinate Systems and Feature Locations

The numbers running along the boundaries of the map refer to three different coordinate systems: Latitude and Longitude, the Universal Transverse Mercator System (UTM), and the State Plane Coordinate System (SPCS).

Latitude and longitude coordinates are printed every 2'30" between the map boundaries. They are marked with black tic marks that are plotted on the inside edge of the map boundaries.

UTM coordinates, on this quadrangle, are marked by the grid. They represent distances in meters and usually are listed using two font sizes. When specified this way, there is an assumed 000 that should be added after the numbers to arrive at the correct coordinate specification. A reading of $_6$**69**, for example, would be reported as 669,000 meters.

The SPCS measures distances in feet, and are readings marked with solid black tic marks that are plotted outside the map boundaries. In this section, we will explore coordinate locations for the *Four Corners Monument,* a marker pinpointing the only U.S. location where the corners of four states – Colorado, New Mexico, Arizona, and Utah – meet.

1. Scroll along the top of the map to find the *Four Corners Monument* – it will be at the top, toward the middle of the quadrangle, where the four states meet, marked by a small symbol at the end of a road as indicated by the red arrow (Figure 5.3)

FIGURE 5.3
Source U.S. Geological Survey, Department of the Interior

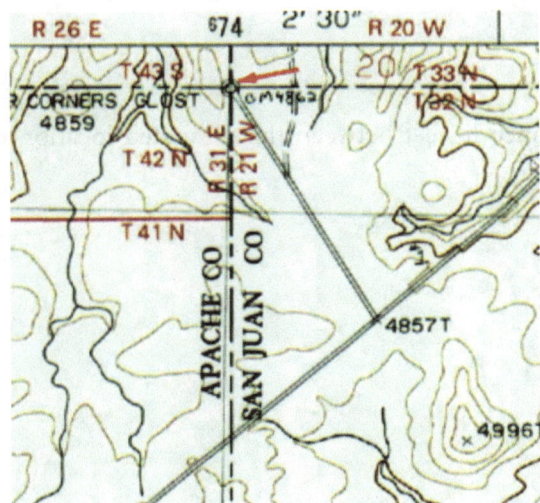

2. Roll your mouse pointer over the location of the monument and note the coordinate locations that are displayed on the bottom right of your ArcMap interface.

QUESTION 5.7

Which of the following coordinates best approximation of the actual location of the monument?

    a. 79° 45' 03" E, 36° 04' 21" N
    b. 79° 45' 03" W, 36° 04' 21" N
    c. 36° 59' 57" E, 109° 02' 43" N
    d. 109° 02' 43" W, 36° 59' 57" N

Establishing the Location of the Four Corners Monument

When first establishing the monument, Congress stated that the monument should be placed at 37° 00' N latitude and at the 32nd meridian of longitude west of the Washington, DC meridian. Using the best available surveying methods and technology in the 1800s, the location of the monument was estimated to be at 37° 00" N, 109° 03' W. However, the monument's actual location versus the estimated location is off by a small distance. See www.ngs.noaa.gov/INFO/fourcorners.shtml for a synopsis of locating the monument.

How far is the actual location off from the estimated location? Let's explore to see.

1. Click on the *XY Tool* located on the *Tools* menu (Top image in Figure 5.4)
2. Type in the intended coordinates (Long: 109 3 0 W and Lat: 37 0 0 N) and press *Add Labeled Point* (bottom image in Figure 5.4)
3. Use the *Measuring Tool* to determine the distance between the estimated location and the actual location of the monument

QUESTION 5.8

What is the approximate distance in feet between the estimated location and the actual location of the monument?

    a. 5,000 feet
    b. 2,500 feet
    c. 1,660 feet
    d. 2,100 feet

FIGURE 5.4

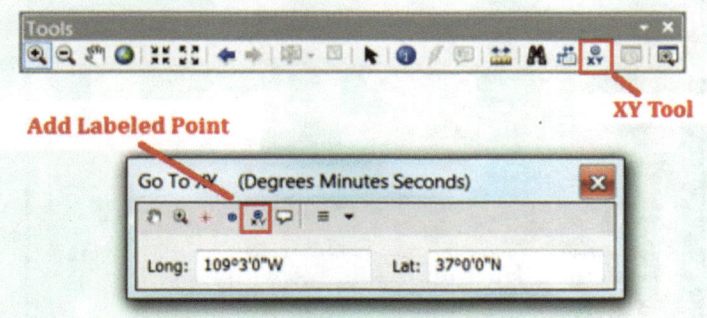

UTM Coordinates

You may not always be in a digital environment that has software tools to automate this kind of measurement task. In that case, it becomes important to have a planar coordinate system, like the UTM system, at your disposal to enhance your ability to complete these tasks manually. Latitude and longitude is a 3D system with coordinates that are measured in angles; in addition, the distance covered by a degree of longitude changes as you move from the equator and towards the poles. The UTM system eliminates these complexities—it is a 2D linear system where distance relationships are constant across the map, making it easier and quicker than latitude and longitude when it comes to calculating distances and directions in the field.

In this section, you will change the display units to UTM coordinates and calculate the distance between the estimated and actual locations of the Four Corners Monument once again except you will be using the *Distance Formula* rather than the *Measure* tool.

1. Right-click the *Data Frame* in your TOC and select *Properties*
2. Under the *General* Tab, change the *Display* to UTM and then click *OK* (Figure 5.5)
3. Move your mouse around the map and be sure the display units are in UTM (units are displayed in the lower right portion of your ArcMap window)

FIGURE 5.5

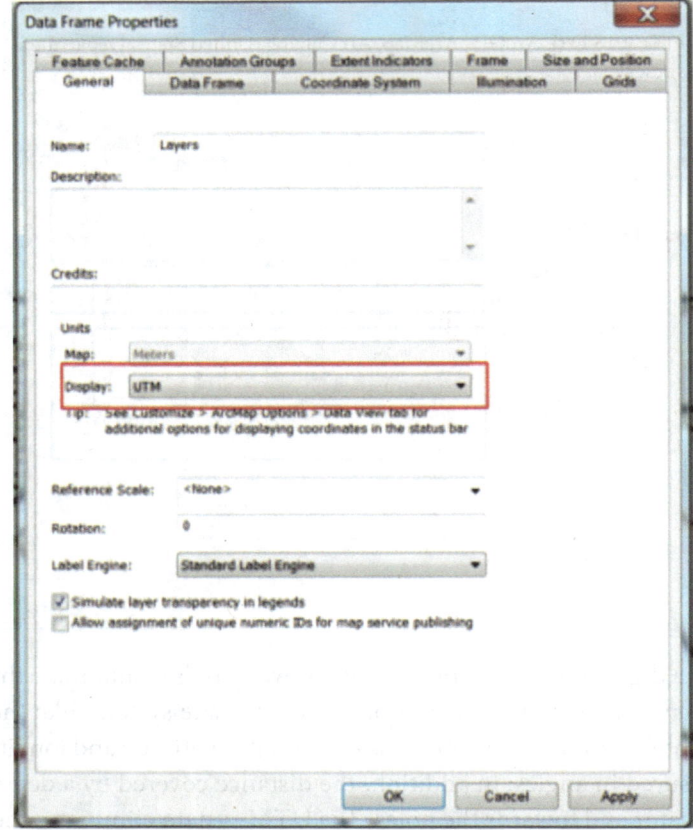

QUESTION 5.9

Which of the following NAD27 UTM coordinates best represents the estimated location of the monument?

a. 650,276 m E; 4,896,428 m N
b. 673,566 m E; 4,096,456 m N
c. 4,896,428 m E; 673,566 m N
d. 4,096,456 m E; 650,276 m N

QUESTION 5.10

Which of the following NAD27 UTM coordinates best represents the actual location of the monument?

a. 674,004 m E; 4,096,347 m N
b. 603,516 m E; 4,287,118 m N
c. 4,287,118 m E; 603,516 m N
d. 4,096,347 m E; 674,004 m N

Computing Distance

With the UTM coordinates we have specified for both the estimated location and the actual map location of the monument, we can assess the actual extent of the displacement by hand to determine the distance between the two locations.

To compute distances, the *Distance Formula* can be used in conjunction with the planar coordinates of the two features of interest. The *Distance Formula* is a derivation of the familiar Pythagorean Theorem, where easting (x) and northing (y) coordinates of the two features are used to determine the distance between them.

QUESTION 5.11

Using the NAD27 UTM coordinates we previously identified (Questions 5.09 and 5.10) and the *Distance Formula*, calculate how far apart the two locations are in meters (be sure to show your work).

499.59

The formula is: $d = \sqrt{(x_2 - x_1)^2 + (y_2 - y_1)^2}$

Where:

x_2 = Easting NAD27 UTM coordinate for the estimated monument location
x_1 = Easting NAD27 UTM coordinate for the actual monument location

y_2 = Northing NAD27 UTM coordinate for the estimated monument location
y_1 = Northing NAD27 UTM coordinate for the actual monument location

Contour Lines

The symbol that sets topographic maps apart from all others is the contour. Brown in color, the contour is an imaginary line that represents locations on a map that share the same elevation value. These lines provide symbology for visualizing 3D changes in terrain by visually interpreting *contour shapes* and *spacings*. Contours that form closed loops, for example, indicate either a summit (hill) or a depression, depending on the type of contour line marking the area. If you see contours that form V's, then you are in a valley with a stream. Closely spaced contours are indicative of steep hills or rugged terrain where change in elevation is occurring rapidly over a small horizontal distance.

The contour interval of the topographic map in Figure 5.6 is 40 feet. Use this map and to answer the following questions (to view the map onscreen, *open* **Contour_Map.pdf** in your Chapter05 folder):

Contour Interval is 40 feet

QUESTION 5.12

Ice L. is lower in elevation than *Lake Calvin*.

a. True
b. False

QUESTION 5.13

If you hiked from *Lake Calvin* to *Lake Margaret* along the A-B transect, or cut line, the steepest part of your hike would be the first half.

 a. True
b. False

QUESTION 5.14

What is the elevation of the highest contour at the summit of Chimney Rock?

a. 11,960
b. 12,160
c. 12,200
 d. 12,640

QUESTION 5.15

The stream connecting Lake Margaret and Elena's Lake is running from Lake Margaret to Elena's Lake.

a. True
b. False

Chapter 6

Thematic Mapping

The same technological innovations that helped drive changes in the accuracy of large-scale topographic mapping also broadened the role of maps in portraying spatial features. The data that explorers collected during their forays in the 17th and 18th centuries included not only locational information, but also information about the characteristics of those locations. Once most of the locational information was ensconced in map form, cartographers and other scientists turned their attention to these characteristics. Their interest in this data was initially driven by a desire to "see" the natural history of the places they had mapped. The realization that the *space* of a location could be mapped as well as its *place* gave us a new form of mapping we know today as the thematic map.

Because of this focus on space, thematic maps are quite different from general reference and topographic maps. Where the general reference map focuses on accurately locating multiple features within a region, the thematic map pushes the locational framework into the background of the map to serve as context, focusing instead on a specific theme or attribute distribution. Early thematic maps focused on portraying the physical characteristics of a place, such as its terrain, vegetation, or climate. With the development of statistics as a discipline in the early 1800s, interest in social or cultural characteristics like population change and incidences of crime and disease also began to be mapped. The point of this new form of mapping was to "...discover the geographical structure of the subject" (Robinson, 1982:16). Today, the themes thematic maps undertake to portray are practically unlimited; they include the traditional physical and population characteristics as well as economic and other social attributes (i.e., religion, income, poverty). Displaying all this information using location as the context required the development of several new mapping techniques, most of which were developed during the early 1800s. These are still widely used today and include such forms as the isoline, the choropleth, the dot, and the proportional symbol.

THE ISOLINE MAP

Isoline maps depict changes in continuous surfaces, such as elevation, temperature, and precipitation, by using line symbols that connect data points with equal values (Figure 6.1). A generic term covering all forms of isarithms, many of these lines also have specific names, depending on the topic mapped. *Iso-* is Greek for *equal*, so an isoline of temperatures is called an *isotherm*. *Isobars* are lines of equal barometric pressure and *isohyets* are lines depicting equal amounts of rainfall. Probably the best-known isoline is the one we explored in Chapter 5, the contour, or *isohypse,* a line of equal elevation.

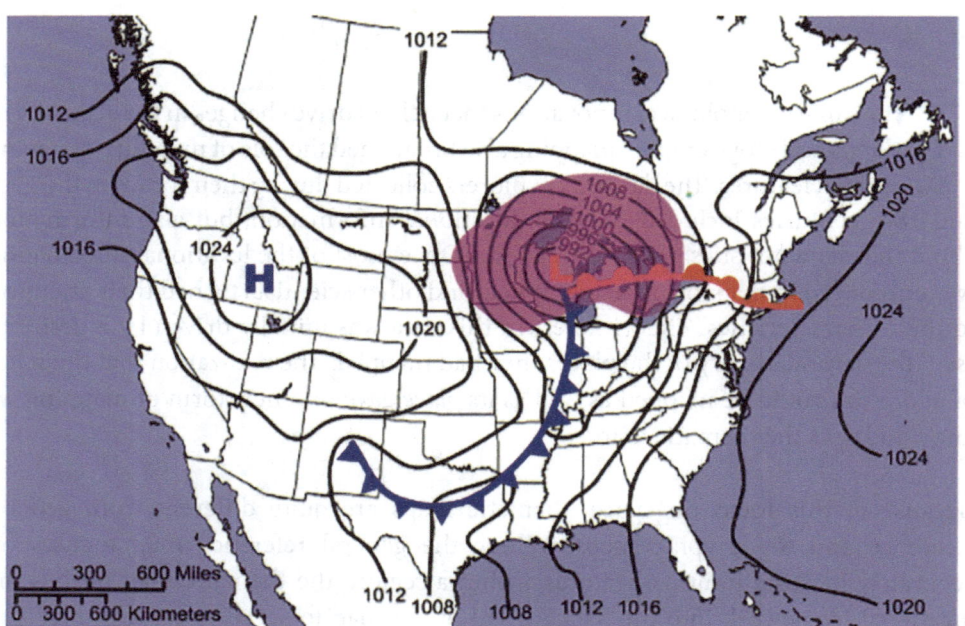

FIGURE 6.1
Mapping wind pressure using isobars. Map: *South Carolina Department of Natural Resources. State Climatology Office.*

The isoline map was the first thematic technique developed. Although used earlier, it was first published as a technique by Edmond Halley in 1701. He used the technique to produce a chart of magnetic variation of the Atlantic Ocean. Lines of equal magnetic variation are called *isogons*; they are also sometimes called *Halleyan lines*, in honor of Halley's part in their development. A British scientist perhaps best known for determining the orbit of Halley's Comet, Halley was also the first to bear the title

EXPLORING GEOSPATIAL TECHNOLOGY

of *thematic cartographer* (Robinson, 1982). His isoline concept was expanded upon in 1817 when Alexander von Humboldt, a Prussian geographer famous for his work in establishing the foundations of biogeography, produced the first isothermal map showing lines of equal temperatures in the Northern Hemisphere.

Today, isolines are used to map a wide variety of topics. These include not only the traditional, physical geography topics, but also cultural and social subjects, like infant mortality rates, disease incidences, and even ancestry patterns inferred from genetic studies. Unlike the more traditional subjects of elevation, temperature, and precipitation, which are portrayed using values that are measured at points across a surface, cultural and social subjects are conceptual and are typically visualized by using data values collected over discrete areas, which decreases the level of accuracy associated with its interpretation. In cartography, these types of maps are termed *isoplethic* (Figure 6.2), as opposed to *isometric*, which denotes those isolines constructed the traditional way.

Regardless of the underlying form, both isoplethic and isometric maps are interpreted in the same manner. The purpose of the display technique is to portray patterns in the data. The distance between two adjacent lines is termed the *isoline interval*; the rate of change between the lines is assumed to be even. Because lines connect points of equal value, each line can have only one value. On one side of the line, data values will be higher; on the other side, they will be lower. The more closely spaced the lines are, the more rapidly the area changes in value.

FIGURE 6.2
Using isopleths to map malaria areas and risks.

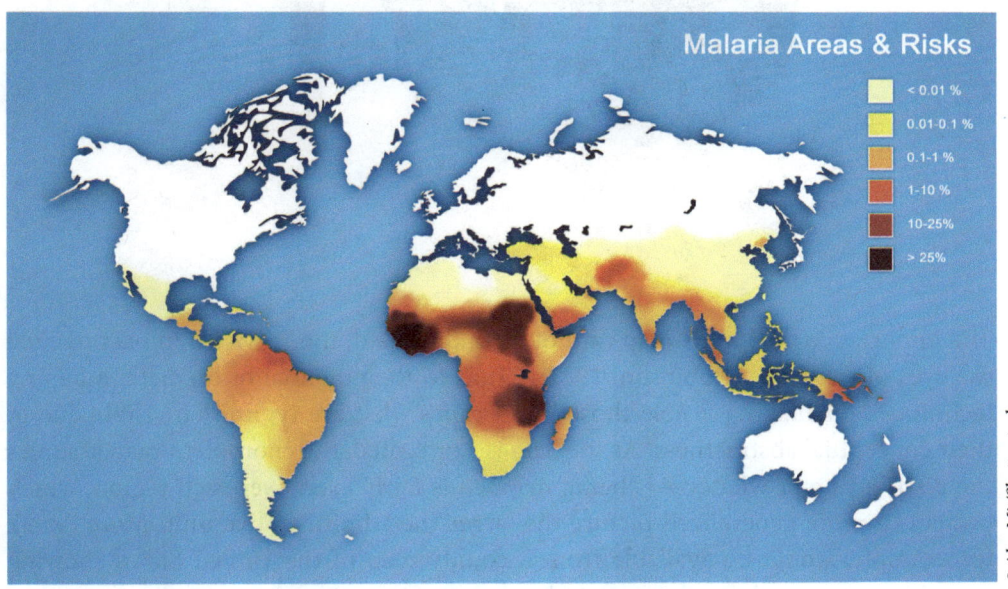

© Alex Mit/Shutterstock.com

The Choropleth Map

While the isometric map design led the way in visualizing distributions of continuous, physical data sets, it was the *choropleth* map, defined as "A method of cartographic representation which employs distinctive shading applied to areas other than those bounded by isolines" (Dent, et al., 2009), that became the standard for mapping data values aggregated over administrative or political regions (Figure 6.3). Baron Charles Dupin, a French mathematician, developed the initial prototype in 1827. Dupin published in several venues, including education and economics, as well as mathematics. One of his many interests was the education of French workers and its relationship to the economy (Delamarre, 1909). He invented the choropleth map to visualize the level of educational instruction in France by shading the different departments of the country according to the number of people per male child in school (Robinson, 1982). Lighter shades represented lower rates of illiteracy (less people per male student) and darker shades represented higher rates of illiteracy.

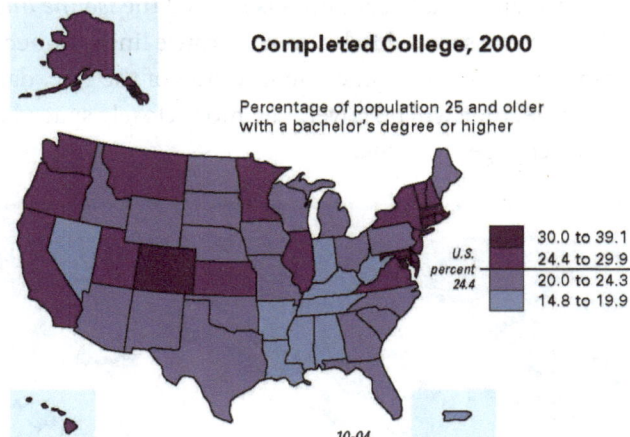

Completed College, 2000

Percentage of population 25 and older
with a bachelor's degree or higher

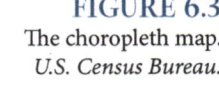

30.0 to 39.1
24.4 to 29.9
20.0 to 24.3
14.8 to 19.9

U.S. percent 24.4

10-04

FIGURE 6.3
The choropleth map.
U.S. Census Bureau.

The importance of Dupin's map cannot be understated; the "...visual impact opened the eyes of scholars" (Robinson, 1982:199) as to the potential of thematic mapping techniques for making visible those "landscapes" that are typically invisible due to their conceptual abstractness. As the data is aggregated, the choropleth map assumes the distribution of whatever is being mapped is fairly even over each region, which presents a more generalized picture the larger the administrative unit. More detail, for example, would be available from a county map of the United States showing variation in some subject than a state-level map of the subject covering the same area; compare Figure 6.4 with Figure 6.3, for example. Whether or not the data is classified prior to symbolization also affects how much detail is embedded in the map. Dupin's prototype did not classify the data, but showed each department's level of illiteracy with a unique shade graded according to its value.

FIGURE 6.4

FIGURE 6.4
Smaller administrative
units give a more
detailed view of patterns
in choropleth mapping.
U.S. Census Bureau.

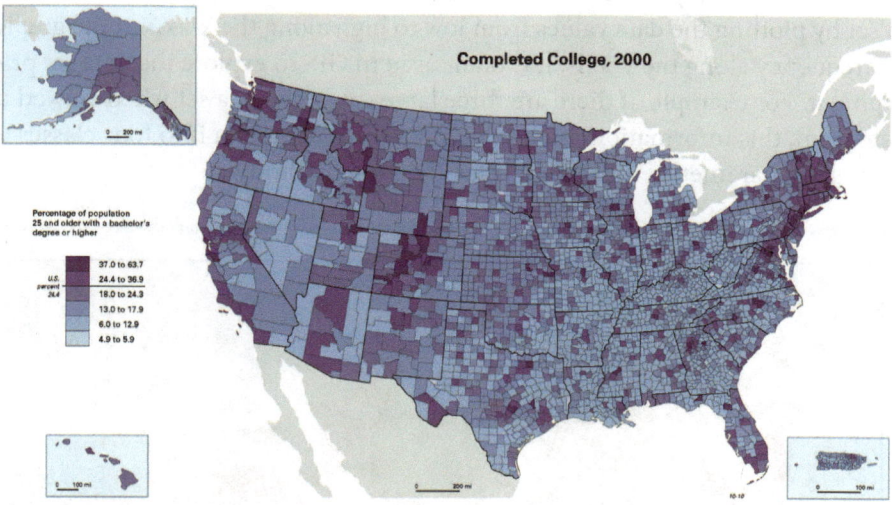

DATA CLASSIFICATION METHODS

The modern choropleth map, due to perceptual limitations in our abilities to discriminate between multiple levels of shading, typically uses classified data sets. The first choropleth map to attempt classification was produced in 1828; it was a Prussian map of population density, but little is known about the designer of its thematic component (Robinson, 1982). Because the choropleth map is one, if not the most popular thematic mapping technique, researchers have put a great deal of effort into perfecting its design. As part of that work, different data classification methods have been developed and studied. The objectives of these methods are to create a classification that 1) is understandable to the map user, and 2) minimizes cartographic bias. Different methods can produce wildly different interpretations of the same data set, which affects the information a map user receives when using the map. GIS will vary in how many, and which, methods are offered, but almost all will provide these: *Natural Breaks*, *Quantile*, and *Equal Interval*.

Natural Breaks. Part of minimizing bias in the map means that from a classification perspective, the least biased interpretation will 1) maximize the numerical differences between class groups, and at the same time, 2) minimize the numerical differences within each of those groups. Of the three methods mentioned, *Natural Breaks* comes the closest to achieving this for any given data set (Figure 6.5). Traditionally, this method works by sorting the data values in the set in ascending order, and then identifying the largest gaps between data values. The gaps become the locations where the class breaks are placed. In most GIS, the manual implementation of this method is called *Manual Breaks;* the Natural Breaks option is automated using the *Jenks Optimization Algorithm*, leaving the cartographer only with the decision of how many classes to use.

The gaps in a data set are also visible when it is graphed using a *histogram* (Figure 6.6). The histogram visually presents the statistical distribution, or underlying form, of the data set by plotting the data values from low to high along the x-axis of the graph and their frequency along the y-axis; it is often used in GIS to explore the data set prior to mapping it. For example, if there are three large gaps in a data set when viewed using a histogram, this information might be used to break the data into four classes when choosing the parameters for the final classification method.

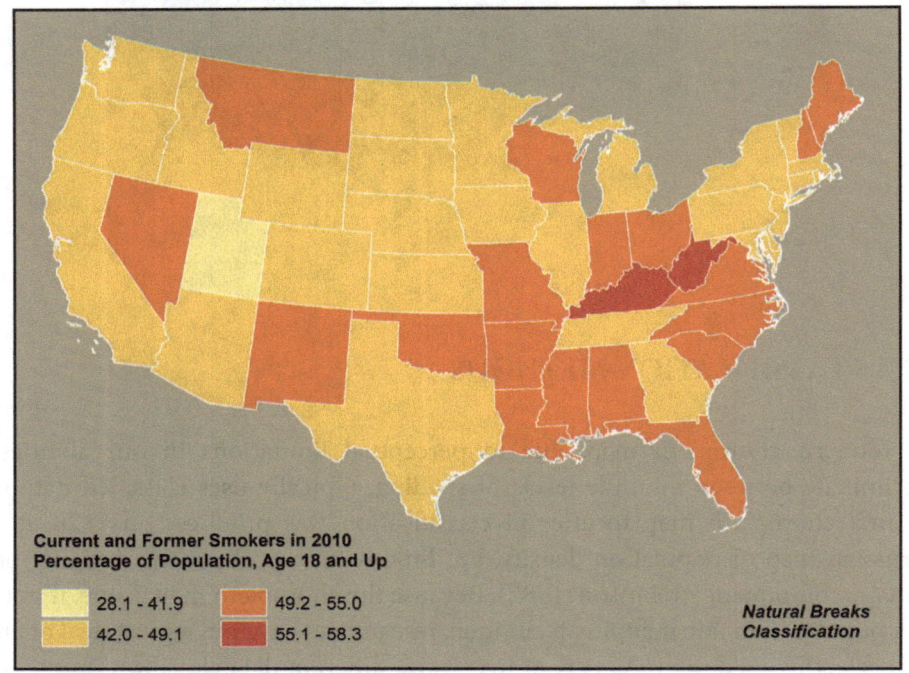

FIGURE 6.5
Percent of Smokers and Ex-smokers age 18 and up, 2010. Distribution classed using Natural Breaks with four classes. Data: *Made with Natural Earth. Free vector and raster map data @ naturalearthdata.com; U.S. Census Bureau.* Source: Elisabeth Nelson

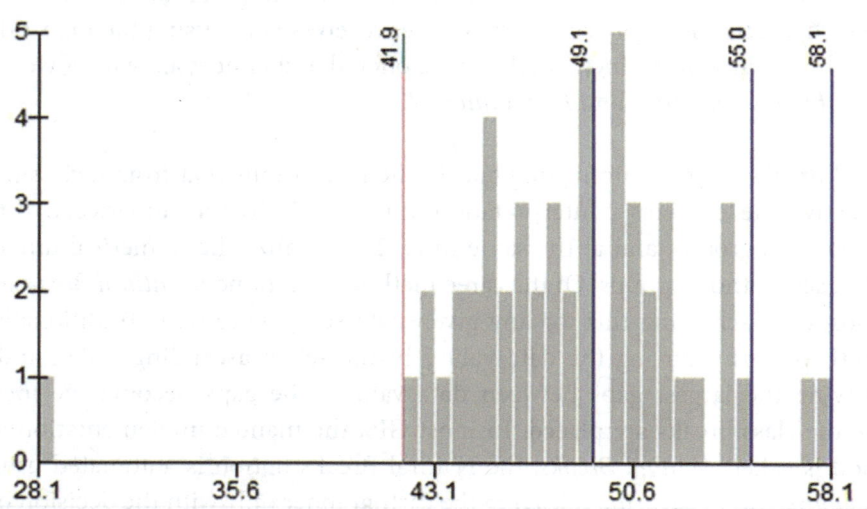

FIGURE 6.6
Histogram of Smoking dataset showing Natural Breaks class limits. Data: *U.S. Census Bureau.* Source: Elisabeth Nelson

In addition to Natural Breaks, the other two most popular classification methods are *Equal Intervals* and *Quantiles*. Both are *blind classification* methods, meaning that they mathematically subdivide the data set into classes without regard to the form of the underlying data. Because of this limitation, they are typically used effectively only with uniform distributions, when making comparisons among multiple data sets, or when required for adhering to special rules aiding interpretation.

Equal Intervals. The Equal Interval classification method divides a dataset into equivalent ranges of data values (Figure 6.7). These ranges are identified by following a series of steps. First, the range of the entire data set (maximum data value – minimum data value) is divided by the number of classes desired. The result is the *numerical width of the range,* which is used to define each individual class range. The lowest class range is then found by adding this number to the lowest data value in the data set to determine the first class break or limit. In Figure 6.7, the numerical width is 7.5, and when this is added to the lowest data value, gives a class break of 35.6 for the lowest data class. As can be seen from the histogram in Figure 6.8, adding 7.5 to 35.6 gives an upper range of 43.1 to the second class; its lower range is the next data value up from 35.6 (35.7). The process continues until all the classes have been defined. If the data are uniformly distributed, then this is an appropriate choice for classification. Unfortunately, most geographical data sets are not uniform distributions, but skewed. When used with these types of statistical distributions, it is important to realize that interpretation can be biased. Because Equal Intervals blindly classifies the data, it is quite possible to have classes in which there are no data values, as well as classes in which dissimilar data values are grouped. On the plus side, this algorithm treats

FIGURE 6.7
Percent of Smokers and Ex-smokers age 18 and up, 2010. Distribution classed using Equal Interval with four classes.
Data: *Made with Natural Earth. Free vector and raster map data @ naturalearthdata.com; U.S. Census Bureau.*
Source: Elisabeth Nelson

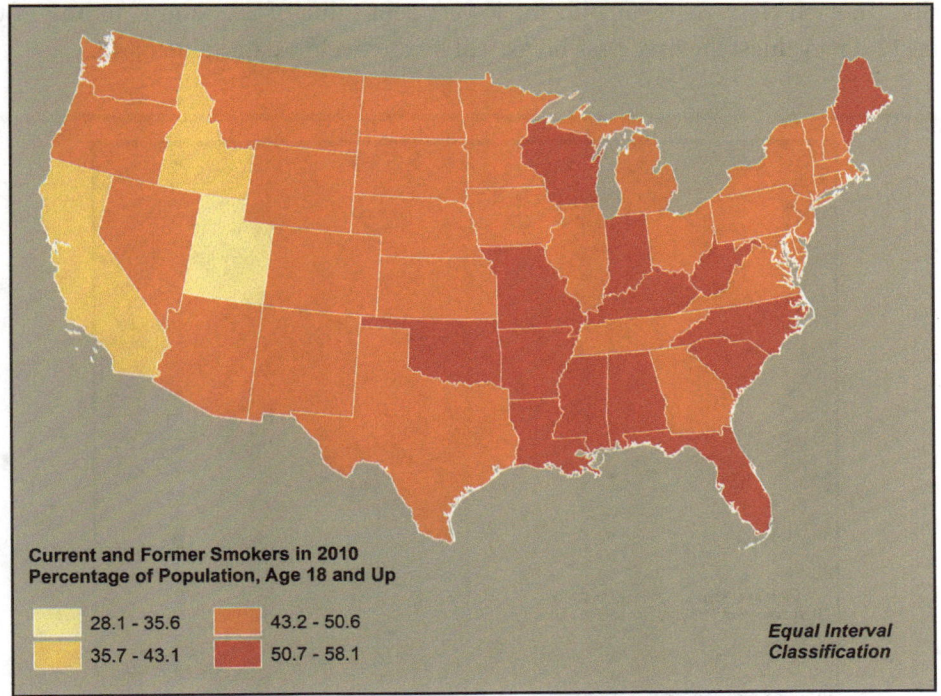

Current and Former Smokers in 2010
Percentage of Population, Age 18 and Up

28.1 - 35.6		43.2 - 50.6	
35.7 - 43.1		50.7 - 58.1	

Equal Interval Classification

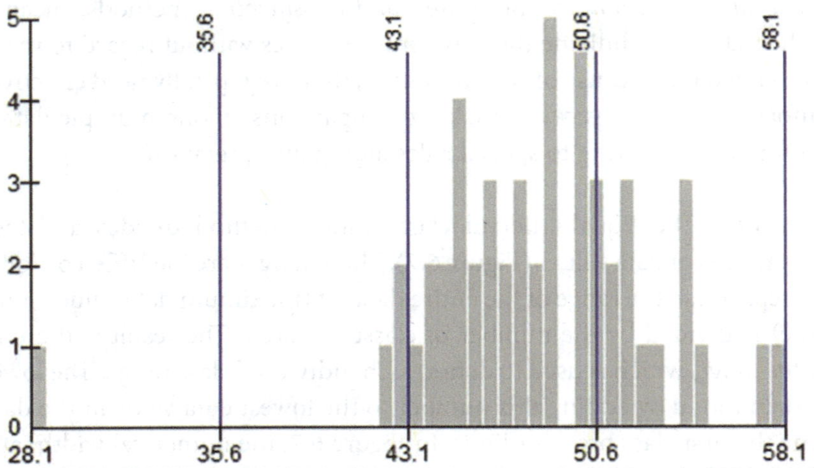

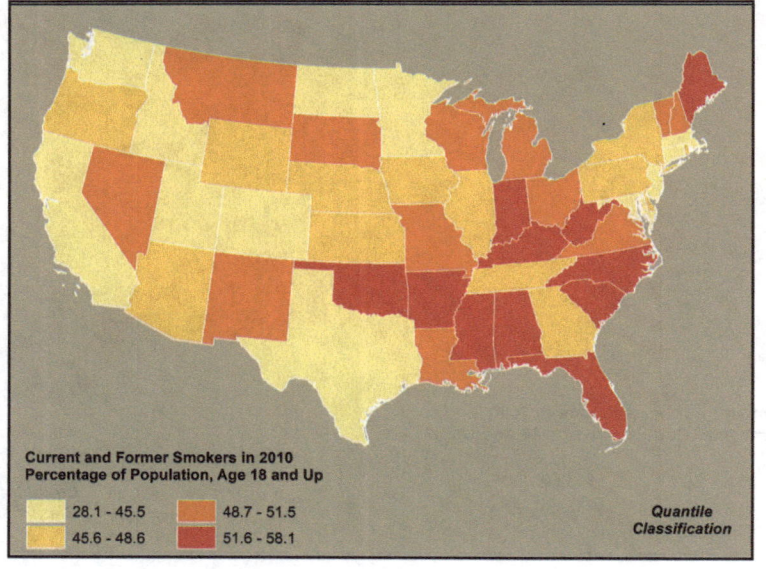

FIGURE 6.8
Histogram of Smoking dataset showing Equal Intervals class limits.
Data: *U.S. Census Bureau*
Source: Elisabeth Nelson

every data set the same way, regardless of distribution, so it is useful when asking the map user to compare two spatial distributions. It is also useful when working with percentage data, as the class widths are always some proportion of 100%, which most map users can easily grasp.

Quantiles. Like Equal Intervals, Quantiles also divides up the range of data indiscriminately, only it assigns an equal number of data values to each class as opposed to an equal range of data values (Figure 6.9). Class breaks are determined by dividing the total number of data values in the data set by the desired number of classes. The result is the number of data values that should be placed in each class. In Figure 6.9, for example, the data set has 48 values and 4 classes, so there should be 12 values in each class. Starting with the lowest data value of 28.1 and counting of the first 12 data values, the first class break will be the twelfth data value or 45.5.

FIGURE 6.9
Percent of Smokers and Ex-smokers age 18 and up, 2010. Distribution classed using Quantile with four classes.
Data: *Made with Natural Earth. Free vector and raster map data @ naturalearthdata.com; U.S. Census Bureau.*
Source: Elisabeth Nelson

The next data value marks the beginning of the second class break, and the same number of data values are counted off, continuing the process until all of the classes have been created. Like Equal Intervals, this is an appropriate choice if the data set is uniformly distributed. If not, then bias can occur, particularly if two like data values are split into two different classes or if classes are created from very dissimilar values. Traditionally considered significant limitations, these potential biases have been shown to be less important than originally thought in one study (Brewer and Pickle, 2002). Using a series of general map reading tasks and epidemiological data, these researchers found that Quantiles outperformed even Natural Breaks for successfully completing tasks. Brewer and Pickle speculate that this is the result of the median value of the data set consistently falling in the middle class, allowing data classes to be approached as ordinal rankings above and below that value; this perspective negates the problem of relative differences in data values within classes. Quantiles also offers one of the most pleasing visualizations, as each class shade is used an equal number of times on the map.

COLOR SCHEMES

Once the data is classified, it is ready to be symbolized and displayed on the map base. For quantitative data sets (ordinal, interval, and ratio levels of data), the tradition is to shade the classes from light to dark, with the lighter shades representing lower data values and darker shades representing higher data values. The idea is to create a visualization that is intuitive, and cartographers have found that most people associate darker, richer colors with 'more' and lighter, less saturated colors with 'less'. The easiest way to implement this is to choose one hue, such as blue, and vary its value from one class to the next to create the shading. This type of color scheme is called a *sequential single hue scheme.*

Sequential multi-hued schemes are also available; these are composed of a transition of adjacent hues from the visible spectrum of light energy. The visible spectrum is the region of light energy that the human eye can process and 'see' as color. The different wavelengths form the rainbow and are ordered according to the acronym ROY G. BIV (red, orange, yellow, green, blue, indigo, violet). The key to creating a successful multi-hued scheme is to start with a light hue, such as yellow, and transition through one or two adjacent and increasingly darker hues, like orange and red, such that the light-to-dark look of the map takes visual precedence over the use of the multiple hues. These schemes, when well designed, offer the advantage of being able to produce more discriminable classes than the single-hue sequential scheme. Generally, the single-hue sequential scheme can support up to six classes before it becomes difficult to tell the shades of a hue apart. Because the sequential multi-hued scheme has two or three hues, the number of classes can be extended to somewhere in the range of eight to ten classes. Increasing the number of classes allows the cartographer to offer more detail about the subject being mapped; as the number of classes decreases, data values are grouped into larger and larger chunks, which obscures some of the finer variations occurring in the data. Many cartographers use a lower limit of four classes because of this, except for special circumstances.

Diverging color schemes are also popular. In these schemes, a light hue represents the middle of the color scheme, with each end diverging into a separate hue to highlight values above and below the midpoint of the data set. A data set, for example, might be represented using white for the class with the midpoint, diverging into shades of red for values above the midpoint and shades of blue for values below the midpoint. This scheme is more specialized; it is typically used with data sets that have been classified around the mean of the data set. It may also, however, be useful for interpreting data classified as quantiles, should the emphasis be placed on the rankings of classes above and below the median value.

UNIQUE DATA REQUIREMENTS

While the choropleth map is the most popular technique for mapping in a GIS, it is not as simple a choice as the default parameters make it appear. Since this technique uses the administrative areas of the base map as the actual symbols of the map, there is no standard size associated with the symbolization; thus, when interpreting the patterns of variation caused by shading them, this lack of consistency can impart misleading information depending on the type of data being mapped. The attribute data symbolized in thematic maps can be grouped into two basic types: *total values* or *derived (ratio) values*. Total values are counts associated with a topic of interest, such as total population or number of influenza cases. These types of data are often significantly influenced by the size of the administrative area from which they are collected, with larger areas home to greater numbers of people. This correlation makes the interpretation of total number patterns suspect when mapped using the choropleth technique. When examining population data for the U.S., as in Figure 6.10, bigger states will often have more people than smaller states; here, five out of the top ten largest states in area are in the top two classes: California, Texas, New Mexico, and Oregon. What most people are associating with a map like this, however, is not the total number of people in a particular location, but how crowded the location is; the only way to get at this interpretation is to process out, or *standardize* or *normalize*, the symbol areas so they do not affect the information being communicated.

The most popular way of standardizing total values for choropleth mapping is to divide each attribute value by the area of the administrative unit; this results in a *density* value. For population data that might be *population per square mile* (Figure 6.11). When treated this way, these data are not visually affected by the different sizes of the areal units. The northeast U.S., for example, is much more densely populated than Texas, but this does not show when mapping total numbers of people. There is also a larger block of more sparsely populated states to the west (Idaho, Montana, Nevada, Utah, the Dakotas, Nebraska, and Kansas) than the map of total population suggests. Data mapped using the choropleth technique, then, should always fall into the *derived values* category. Density measures are the most common, but areas can also be standardized in other ways. For crime data, for example, total values for each area could be divided by the total number of crimes in the country to arrive at a percentage of total crimes for each areal unit. A crime rate could also be created to standardize the data by dividing the total crime data for each area by the total population for each area. If it's total data that you need to map, there are other, more appropriate mapping techniques.

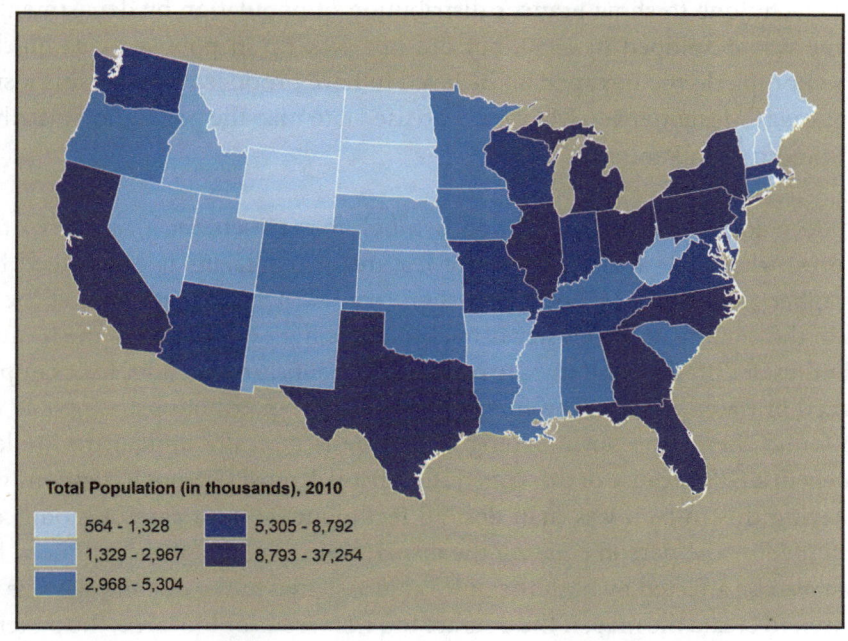

FIGURE 6.10
Total Population in the United States, 2010.
Data: *Made with Natural Earth. Free vector and raster map data @ naturalearthdata.com; U.S. Census Bureau.*
Source: Elisabeth Nelson

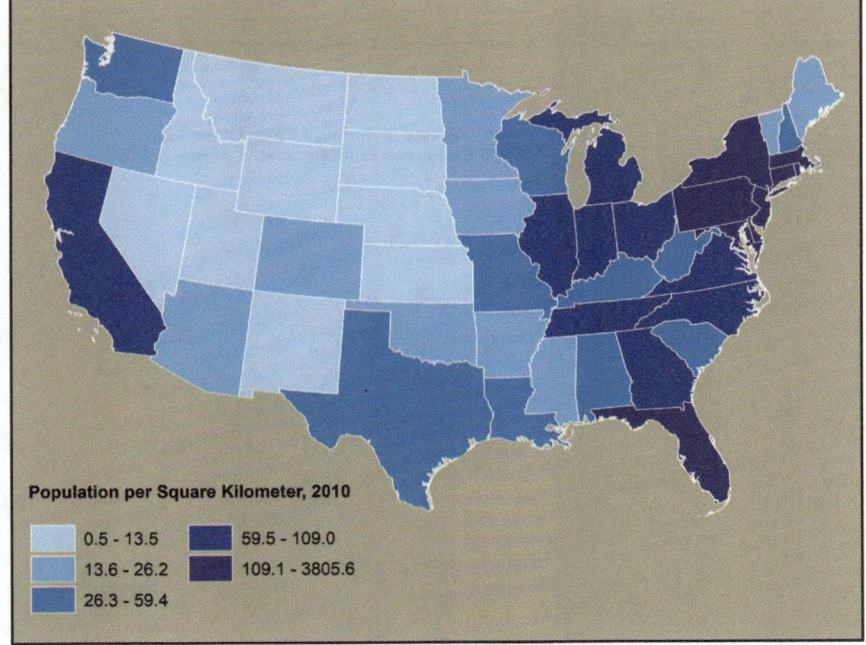

FIGURE 6.11
Population per Square Kilometer, 2010.
Data: *Made with Natural Earth. Free vector and raster map data @ naturalearthdata.com; U.S. Census Bureau.*
Source: Elisabeth Nelson

The Dot Density Map

Perhaps the purest form for mapping total values is the dot density map. Dot maps show the presence of features by mapping them with dot symbols. Like the choropleth map type, the dot map was also first produced in France, where Frère de Montizon used the technique to show France's distribution of population by department. This prototype was developed in 1830, but did not pick up in popularity as quickly as Dupin's choropleth; the next person believed to have produced a map using a similar technique was Alexander von Mentzer, who used it to map the population distribution of the Scandinavian Peninsula.

The modern dot map technique uses dot symbols to represent the location of a cluster of features, where the actual number of features it represents is established by the cartographer (Figure 6.12). Locations, then, are not actual locations, but locations placed in the vicinity of the cluster using other ancillary data, such as the location of water bodies or different land uses, to refine placements further. Dots, for example, are not placed in water bodies when mapping population distributions, as people would not be found there. The dot mapping technique is typically difficult to implement accurately in a GIS because of this constraint. Instead, most GIS opt to randomly place dots, leaving this to be a less than desired technique without careful modifications implemented by the person creating the display. The spatial pattern produced by the dots can also be affected by both the unit value assigned to the dot and the size of the dot chosen to create the map; if the dots are too numerous or too large, the pattern can look more dense than it actually is, for example. On the other hand, dots that are too few or too small may give the inappropriate impression of sparseness.

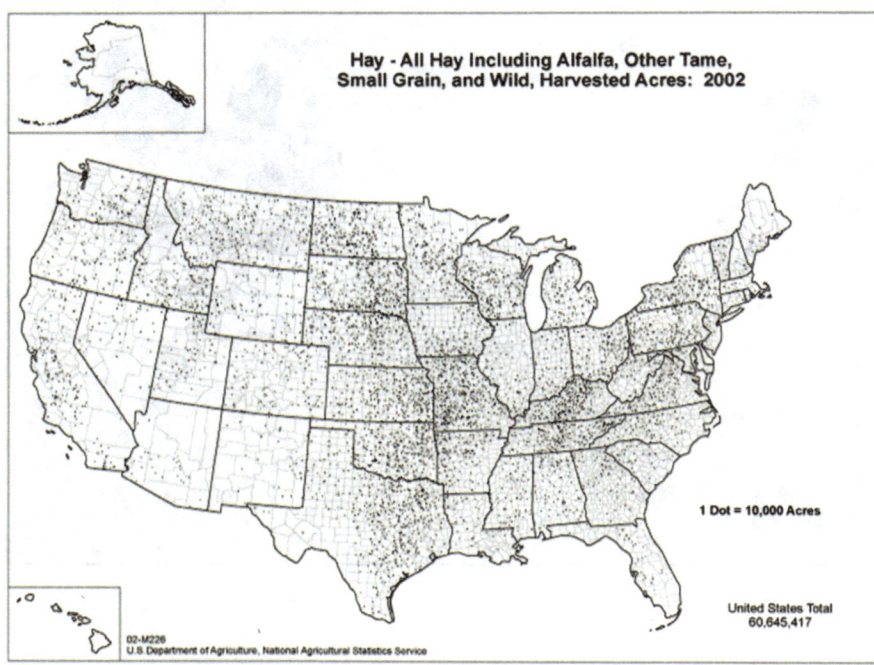

FIGURE 6.12
Dot map showing harvested acres of hay in 2002. *U.S. Geological Survey, Department of the Interior.*

EXPLORING GEOSPATIAL TECHNOLOGY

Proportional and Graduated Symbol Maps

PROPORTIONAL SYMBOL MAPS

Proportional symbol and graduated symbol maps are two other techniques used to map total values, as well as some types of derived values. Cartographic historians believe the first use of these symbols occurred in 1837, when Henry Drury Harness used them to create a map of city populations for the Irish Railway Commission (MacEachren, 1979). In this map, the symbols were circles sized proportionally to the population of the city represented. Proportional symbols can be used to symbolize both discrete point data, like cities, and attribute values aggregated by administrative area, such as county population (Figure 6.13). The technique supports the mapping of total data values, as well as percentages, averages, and rates. Density values, like population per square mile, are traditionally avoided, as the choropleth technique is particularly suited for mapping this type of data.

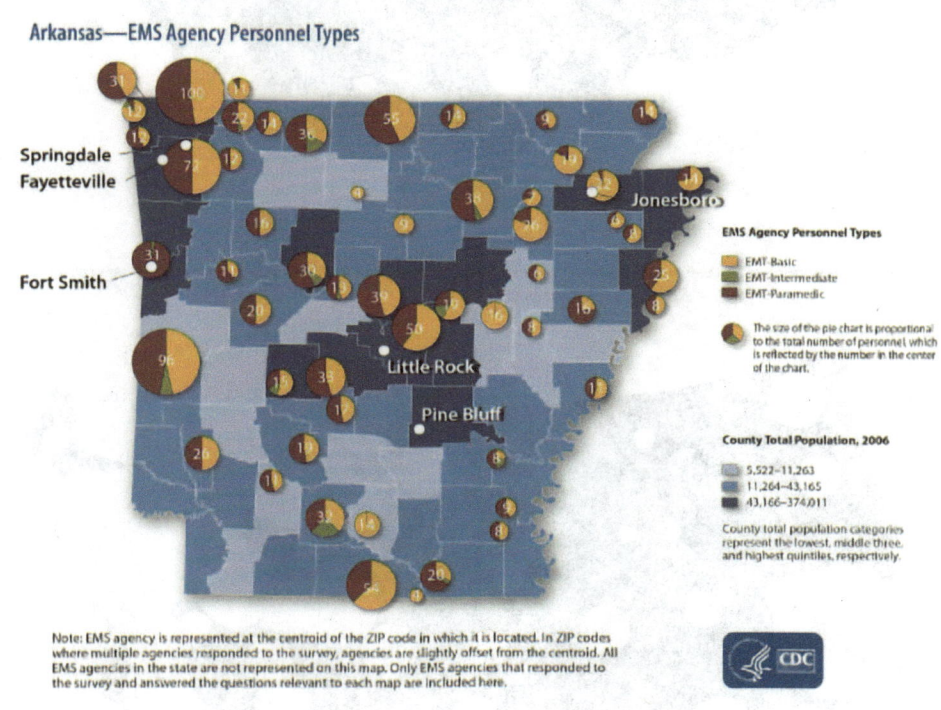

FIGURE 6.13
Proportional symbol map of EMS personnel types in Arkansas. *Center for Disease Control.*

The symbols representing a data set are scaled relative to an *anchor value*. Usually this value is either the smallest or largest data value in the set. A circle size is chosen to represent this data value, and then all other circles are scaled in proportion relative to this value. When choosing an anchor value, it is important that the smallest circles be clearly visible, while at the same time the largest circles are not too dominant visually.

Because of this, one of the more vexing issues with proportional symbols is dealing with the range of data values in a data set. If the range is too narrow, the scaled symbol sizes will appear too similar, making the final display of patterns uninteresting (Figure 6.14). On the other hand, if the range is too large, the largest symbols will be too large to be accommodated by the map if the smallest circle is visible (Figure 6.15). In this case, the total values are often converted to derived values to pull the range in a bit.

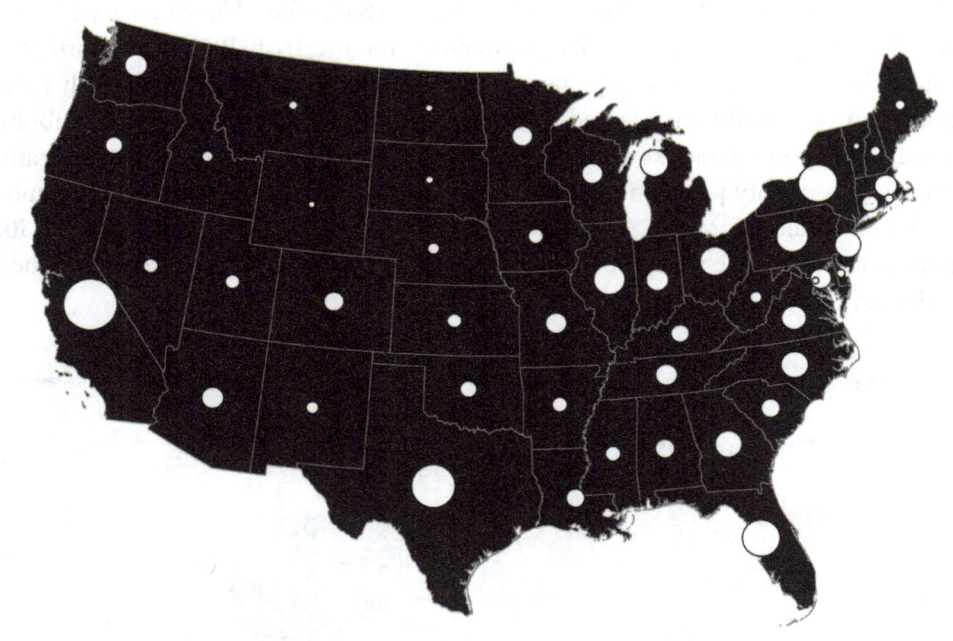

FIGURE 6.14
Proportional symbol map with symbol size range too small.
Data: *Made with Natural Earth. Free vector and raster map data @ naturalearthdata.com; U.S. Census Bureau.*
Source: Elisabeth Nelson

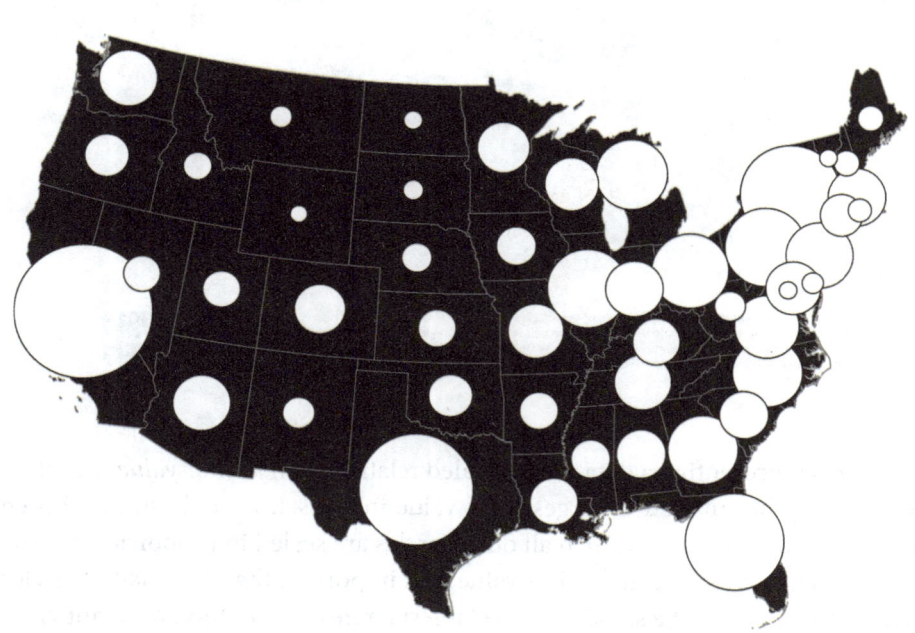

FIGURE 6.15
Proportional symbol map with symbol size range too large.
Data: *Made with Natural Earth. Free vector and raster map data @ naturalearthdata.com; U.S. Census Bureau.*
Source: Elisabeth Nelson

Research has shown that map users have a tendency to underestimate the sizes of the larger symbols in map reading tasks. This is a perceptual limitation on our part; as the dimensions by which a symbol is scaled (length, area, volume, etc.) gets larger, we become increasingly poorer at judging absolute size or relative changes in the symbols. Circles are often the symbol of choice because 1) they "grow" in size more slowly than changes in length, as the growth is spread over two dimensions, 2) they are more easily assessed than changes in volume (3D), and 3) they are a visually pleasing symbol lacking harsh lines that might clash with the base data. Some GIS do offer a *perceptual scaling* option that when checked will rescale the larger symbols to compensate for our perceptual limitations.

GRADUATED SYMBOL MAPS

Graduated symbols maps are a variant of proportional symbols in which the attribute data is classified prior to symbolization. Once the data is aggregated into a set number of classes, a symbol size is assigned to each class of data values (Figure 6.16). These symbols can be scaled proportionally to the mean of each class or they can be chosen so that each size is easily discriminable on the map. Classifying the data prior to symbol assignment reduces the number of differently sized symbols the map user has to process when interpreting spatial patterns. This simplifies the patterns in the data, although the generalization does mean some detail in the data will be lost. This variant of the proportional symbol map also helps control issues with data ranges that are too narrow or too wide by allowing the cartographer a little more leeway in size choices.

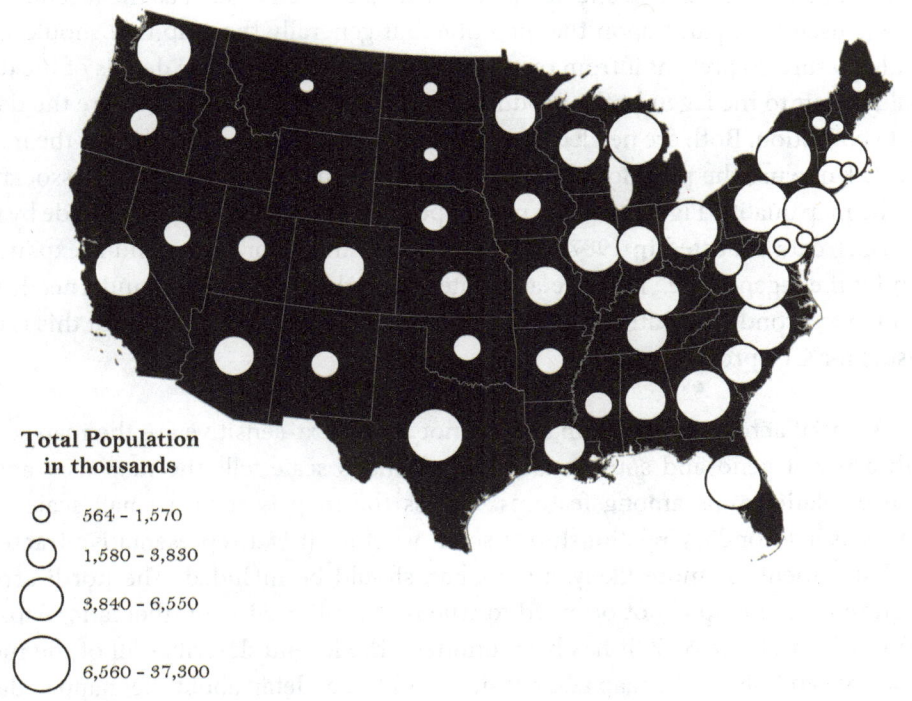

FIGURE 6.16
Graduated symbol map.
Data: *Made with Natural Earth. Free vector and raster map data @ naturalearthdata.com;* U.S. Census Bureau.
Source: Elisabeth Nelson

Total Population in thousands

○ 564 – 1,570

○ 1,580 – 3,830

○ 3,840 – 6,550

○ 6,560 – 37,300

Map Design

The map itself, although by far the most important element, is just one of many in a final cartographic layout. Without surrounding elements, like a title and a legend, to provide additional context, it can be difficult to understand the message the map is trying to communicate. In addition to interpretative difficulties, an "…ugly map, with crude colors, careless linework, and disagreeable, poorly arranged lettering may be intrinsically as accurate as a beautiful map, but it is less likely to inspire confidence" (Wright, 1942:23). Map design addresses these issues and more; it is a complex process involving a series of decisions, each influencing the other. There are guidelines to assist the cartographer – Dent, et al. (2009) lists simplicity, functional appropriateness, pleasing appearance, and economic considerations as some of the primary principles under which cartographers operate. One way to approach map design and implement these principles is to identify the major elements of most maps, and then discuss how they are most effectively used in the layout process by dividing the design structure of the map into issues of *visual balance* and issues of *visual hierarchy*.

MAP ELEMENTS

Foote and Crum (2000), writing for *The Geographer's Craft*, divide map elements (excluding the mapped area itself) into three basic groups: those found on most, if not all maps, those that are essential but sensitive to context, and those used to enhance communication. Essential elements that tend to be context-sensitive include the map title, date of map production, and cartographer credits. Of these, the map title is ranked the most important; it describes the theme of the map and orients the map user. Its content can be affected by other map elements, such as the legend title, which is used to expand upon the map title, but generally the map title should give the 'big picture' to prevent it from overwhelming the map user; the details of the data should be left to the legend title. Production dates are time-sensitive, as are the dates of data collection. Both are needed by the map user to assess the currency of the map's subject. Knowing the producer of the map imparts the level of authority associated with the map quality. Figure 6.17 is a map of population density of Poland made by the CIA using data collected in 1987. The title is short and informative and is expanded upon by the legend title. The date associated with the data is listed underneath the legend in a secondary location; the cartographic agency is not credited, but this is not unusual for CIA products.

Also essential across almost all maps, but not as context-sensitive are the map scale, north arrow, legend and source statement. A map's scale tells the map user about distance relationships among features. Unless the map is such a small scale that distance is a secondary relationship, a scale, whether it is a representative fraction, verbal statement, or more likely, a scale bar, should be included. The north arrow is required if the map is not oriented to true north; otherwise, on thematic maps, it is optional. In Figure 6.17, it has been omitted. The legend describes all of the most important symbols on the map and provides additional detail about the mapped data.

FIGURE 6.17
Map elements.
Annotated map: *Central Intelligence Agency.*
Source: Elisabeth Nelson

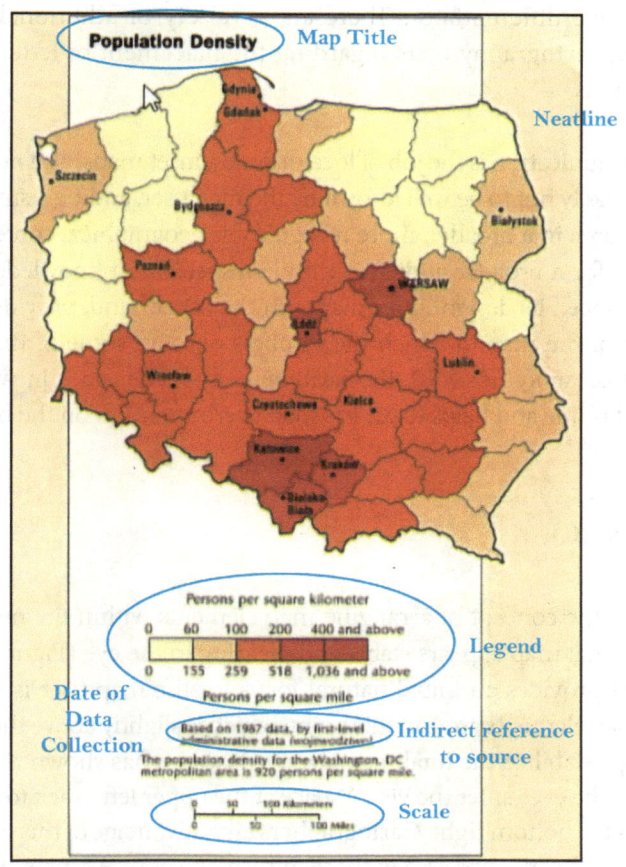

The symbols in the legend should match those on the map exactly and should be described clearly and accurately for the map user. The source statement tells the map user where your data originated and in what form, so they can backtrack and verify your interpretation of the data if so desired. The source in Figure 6.17 is alluded to, but is not specified.

Although not specifically mentioned by Foote and Crum, it would be remiss not to at least mention the importance of labels/text for providing context and additional detail. Many of the elements listed so far do fall under this category, but text and labels comprise much more, from identifying components of the mapped area, like cities, rivers, and countries, to providing interpretive details in block form. Text is such an integral part of mapping that cartographers treat it as a cartographic symbol, using variations in typeface, type size, and type form to create differences among mapped features. Typefaces are families of fonts, and they come in two broad categories. *Serif* faces have small finishing strokes on the ends of the letters (like Harrington or Times New Roman). Sans serif fonts (like Arial or Verdana) do not have finishing strokes. This characteristic of type is often used to make qualitative distinctions between mapped features, such as physical versus cultural on reference maps. Type form includes variations in weight (**bold**) and style (*italic*) and type size, is how large the font is. Weight and size are often used to emphasize important features; style is another qualitative differentiator. There are a variety of additional guidelines that can be found in cartography texts regarding the placement of lettering for different mapping situations.

Enhancing communication is the job of locator maps, inset maps, and neatlines. For maps of areas that are likely not to be well known by the map user, adding a smaller locator map places the main area in a broader, more recognizable geographical context. If Figure 6.17 had been meant for a broader audience with less specialized knowledge, a locator map would have enhanced its design. Inset maps, on the other hand, provide a more detailed view of an area of the map in which the symbols obscure some of the additional map detail. Last, but certainly not least, the neatline defines the space in which all the map elements should reside and helps focus the map user's attention on the map elements.

VISUAL BALANCE

Visual balance is the concept of arranging map elements within the neatline so that the structure of the final map appears stable and pleasing to the eye (Figure 6.18). The point on the page that provides an initial natural focus for the map user is the *optical center*. Balancing the map elements around this point, which is slightly above the *geometric center* of the page, helps stabilize the final map layout. Research has shown that in left-to-right reading cultures, the eyes enter the visual space at the upper left, track to the optical center and exit towards the bottom right. Cartographers take advantage of this pattern to enhance stability by placing the mapped area in the vicinity of the optical center, then arranging the remaining map elements so that the highest ranking elements fall within that path; the less important elements are placed off the path in secondary spaces.

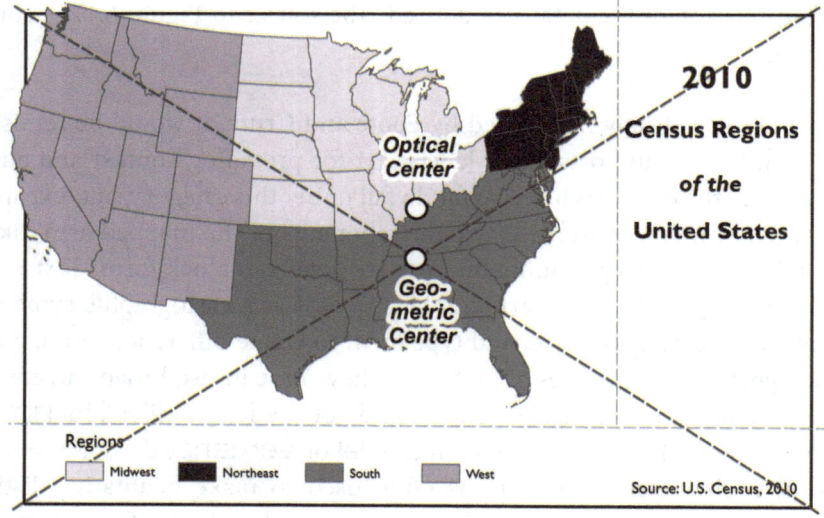

FIGURE 6.18
Implementing concepts of visual balance. Data: *Made with Natural Earth. Free vector and raster map data @ naturalearthdata.com; U.S. Census Bureau.* Source: Elisabeth Nelson

The divisions of the image space, as well as the visual weight of the individual map elements, also play roles in balancing a map layout. As the map elements are being arranged, for example, dividing the space into unequal parts as opposed to a more even and symmetrical design will add visual interest and imbue the space with life (Dent, et al., 2009). The visual weights of the map elements, when considered in relation to one another, add yet another dimension that can affect balance (Arnheim, 1965). Larger objects, for instance, tend to be visually 'heavier' and claim more attention, so they should be balanced against smaller objects. Objects to the right of center, as well as those towards the upper part of the layout are likely to appear heavier, or draw more attention, than those on the left or those positioned towards the bottom of the layout. White space, which most treat as a negative quality, also affects final stability and should be considered using the same approaches as applied to map elements.

VISUAL HIERARCHY

The elements of a map are arranged both on a single visual plane, as with the concept of balance, and from the perspective of an intellectual hierarchy of planes. For any given map, each map element will have a relative intellectual rank that corresponds to the importance of the element in communicating the information presented by the map. Elements with higher ranks are more important to the "story", so their visual importance relative to all the other map elements should reflect that. To reflect different levels of importance in this manner, designers use the concepts of *figure-ground* and *contrast*. Figure-ground is a perceptual tendency we have to categorize what we are seeing into two levels: figures, which are more important and tend to stand out visually, and grounds, which are less important and tend to form a background on which the figures rest. Figure-ground relationships between objects can be enhanced by using *contrast* in the design of map elements. Contrast is the property of an object that makes it distinguishable from other objects. In the visual world, it could be a contrast in weight or detail of an object, such as line thickness, or label detail, or in any of the basic visual variables used in map design: hue, value, saturation, size, shape, orientation, or texture. Varying these properties creates visual interest and improves our ability to visually organize objects into more important and less important objects. Thicker lines, for instance, are usually seen as more important than thin ones; areas with more detail attract more attention than those with fewer details (Dent, et al., 2009). In the realm of color, brighter hues seem to attract more attention than dark hues, and in value, darker values trump lighter ones in visual importance.

Additional References

Andrienko, G., Andrienko, N., and A. Savinov. *Choropleth Maps: Classification Revisited.* http://geoanalytics.net/and/papers/ica01.pdf. Last accessed July 8, 2013.

Arnheim, R. 1965. *Art and Visual Perception.* Berkeley: University of California Press.

Barnes, J. 1978. *Control Areas and Control Points in Isopleth Mapping.* The American Cartographer, 5:1 (65-69).

Bradley, M. *Charles Dupin and His Influence on France.* http://www.cambriapress.com/cambriapress.cfm?template=4&bid=467 Last accessed July 7, 2013.

Brewer, C. and L. Pickle. 2002. *Evaluation of Methods for Classifying Epidemiological Data on Choropleth Maps in Series.* Annals of the Association of American Geographers. v.92(4): 662-681.

Campbell, J. (2001). Map use and analysis (4th ed.). Boston: McGraw-Hill.

Chang, *Introduction to Geographic Information Systems*, 5th ed. McGraw Hill Companies, Inc.: New York, New York.

Delamarre, L. (1909). *Pierre-Charles-François Dupin.* In The Catholic Encyclopedia. New York: Robert Appleton Company. Retrieved July 6, 2013 from New Advent: http://www.newadvent.org/cathen/05205a.htm

Dent, B., Torguson, J., & Hodler, T. (2009). Cartography: Thematic map design. Boston: McGraw-Hill.

Dramowicz, K., and E. Dramowicz. 2004. *Choropleth Mapping with Exploratory Data Analysis.* Directions Magazine. December 29. http://www.directionsmag.com/ Last accessed July 3, 2013.

Friendly, M., and D. Denis. *Milestones in the history of thematic cartography, statistical graphics, and data visualization.* http://datavis.ca/milestones/index.php?group=1800%2B&mid=ms105 Last accessed July 5, 2013.

Foote, K. and S. Crum. *Cartographic Communication.* http://www.colorado.edu/geography/gcraft/notes/cartocom/elements.html Last accessed July 10, 2013

NCGIA. *Data Classification.* http://www.ncgia.ucsb.edu/cctp/units/unit47/html/comp_class.html Last accessed July 7, 2013.

Robinson, A. 1982. *Early Thematic Mapping in the History of Cartography.* Chicago: University of Chicago Press.

Slocum, T., McMaster, R., Kessler, F., & Howard, H. (2009). Thematic cartography and geovisualization, 3rd ed. Upper Saddle River, NJ: Pearson Prentice Hall.

Warf, B. 2010. *Encyclopedia of Geography.* SAGE Publications, Inc.

Wright, J. 1942. *Mapmakers are human.* Geographical Review. v.32(4):527-544.

Exercise 6

Producing a Choropleth Map

In Exercise 6, you will create a choropleth map design and layout that highlights population change in the southeast United States from 2000 to 2010.

The U.S. Department of the Census reported an increase of over 9% in the total population of the U.S, from 281.4 million in 2000 to 308.7 million in 2010. You have been asked to create a choropleth map highlighting changes in the southern census region.

OBJECTIVES

- Set up a basemap
- Design a choropleth overlay by standardizing and classifying attribute data and applying the appropriate color schemes
- Design a map layout

SOFTWARE INFORMATION

Introducing ArcMap uses Esri's **ArcGIS 10.6** software. **ArcGIS** is a commercial GIS package, available in geospatial labs affiliated with schools that have a campus-wide site license for the software. Instructors at these campuses may also request 1-year student versions of the software at http://www.esri.com/landing-pages/education-promo.

DATA

The data for the exercises is available from the Kendall Hunt Student Ancillary site. See the inside front cover for access information. You may also be directed to download the data from a different location by your instructor.

1. Download the data as instructed by your instructor

2. Save the file to a location where you have read/write privileges (USB key, home computer, class server)

The file you just saved is in a compressed (*.zip) format, and was created using **7-zip** freeware (http://www.7- zip.org). To use the data for this exercise, you must decompress it using the same software.

3. Locate the zipped file that you saved
4. Right-click the file
5. Choose *7-Zip - Extract Files*
6. Click *OK* to create a folder with the decompressed data

Specify a Home Folder

In this exercise, rather than opening an already existing ArcMap map document, you will be creating one from scratch. Because of this, you will also need to explicitly associate a Home Folder and a default geodatabase for your new map. This requires a few steps up front, but will make things easier later.

1. Start **ArcMap** by double-clicking on your desktop icon or by clicking >All Programs > ArcGIS > **ArcMap 10.6**
2. In the *ArcMap - Getting Started* dialog box, click *Cancel*
3. If the *Catalog* window is not open, click on the *Catalog* tab on the right side of the ArcMap interface OR go to *Windows – Catalog* to open it
4. Click the pushpin in the upper right corner of the *Catalog* window to freeze the window in place (Figure 6.1)

FIGURE 6.1
Used with permission. Copyright © 2018 Esri, ArcGIS, ArcMap, Arc-Catalog, United States Department of the Census; naturalearthdata.com. All rights reserved.

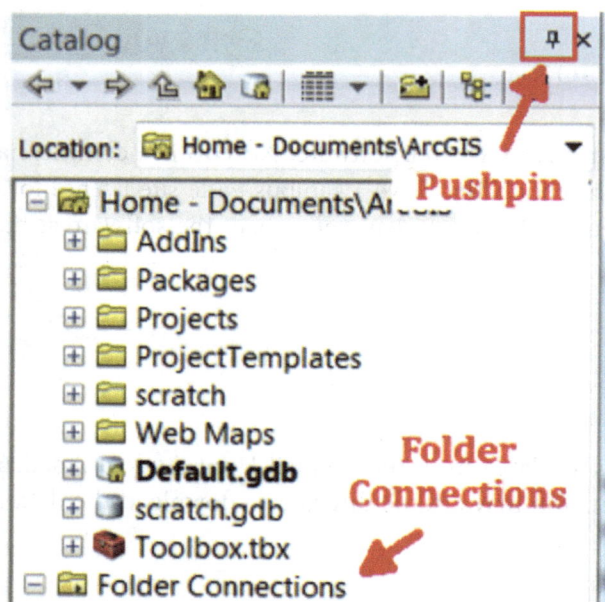

Notice that the default Home folder is set *Home-Documents\ArcGIS*. You want your Home folder to be *Chapter 06*, which is where your data for this exercise resides.

5. From File, choose *Save As*
6. Name your file using this convention: *lastname_Exercise06*
7. Save your map to the *Chapter06* folder containing *Exercise_06.gdb*

Your home folder should now be updated to the correct location. If you have any questions, verify this with your instructor before proceeding.

Specify a Default Geodatabase

It is also important to link your map document specifically to the geodatabase that houses your data for the exercise. With this link in place, any changes or additions you make to your data will be stored in the correct location on the computer.

1. From the *File* menu, choose *Map Document Properties*
2. Click to select the file folder button to the right of the default geodatabase path
3. Browse to find *Exercise_06.gdb*
4. Click to select it, then click *Add*
5. Next to pathnames, place a checkmark by *Store relative pathnames to data sources*
6. Click *OK*

Add Spatial Data to the Map Document

1. In the Catalog, find your *Home* folder
2. Click the + sign to the left of *Exercise_06.gdb* to see the spatial data files you will use in this exercise
3. Click and drag *Census_2000_2010* from the geodatabase to your *Data View* window

Set up the Basemap

Thematic maps almost always require an equal-area map projection for the basemap because users will be comparing values between different areas on the map.

1. Double-click the *Layers* data frame in the *TOC*
2. Click the *Coordinate System* tab to select it
3. In the top scroll box, scroll to find the *Projected Coordinate Systems* folder
4. Click the + sign next to the folder to expand it and view its contents
5. Find the *Continental* folder and expand its contents
6. Find the *North America* folder and expand its contents
7. Choose an equal area projection designed for the United States and click *OK*

QUESTION 6.1

Which of the following map projections is NOT equal-area?
Hint: You can use the Help function in ArcMap to learn more about your choices...

 a. Albers Conic
 b. Behrmann Cylindrical
 c. Mercator
 d. Mollweide

Once projected, the map may appear "skewed" or tilted and need to be re-centered on the region.

8. Right-click the data frame and select *Properties*
9. Click the *Coordinate System* tab
10. In the top scroll box, scroll to find the *selected projection*
11. Right-click the projection and choose *Copy and Modify*
12. Change the *central meridian* value to *-90*
13. Click *OK*, then *OK* again

Save Your Map Document

1. From the *File* menu, choose *Map Document Properties*
2. Click the file folder button to the right of the default geodatabase path and browse to the *Exercise_06.gdb*
3. Click to select it, then click *Add*
4. Next to pathnames, place a checkmark by *Store relative pathnames to data sources*
5. Click *OK*
6. From *File*, choose *Save As*
7. Name your file using this convention: *lastname_Exercise06*
8. Save your map to the *Chapter06* folder
9. Double-click the *Computer* icon your desktop
10. Navigate to the location where you think you saved your file and verify that it is there

Design a Choropleth Overlay

CHOOSE YOUR INITIAL ATTRIBUTE DATA SET

There are several fields in the attribute table associated with population:

Attribute Field	Field Description
POP2000	Total population, year 2000
POP2001	Total population, year 2001
POP_2010	Total population, year 2010
POP00_SQMI	Population per square mile, year 2000
POP_CHG	Change in total population
PPOP_CHG	Percent change in total population

Your map will highlight changes that occurred between the 2000 and 2010 censuses. A field for change in total population has been created, but has yet to be populated with any data – this is the place to start.

1. Right-click *Census_South_2000_2010* and choose *Open Attribute Table*
2. To calculate the raw change in population between 2000 and 2010, right-click the field *POP_CHG* and choose *Field Calculator*
3. Click *OK* to dismiss the warning dialog box if it appears
4. In the bottom window labeled POP_CHG =, type **[POP_2010] - [POP2000]** and click *OK* (Figure 6.1)

STANDARDIZE YOUR ATTRIBUTE DATA

Even at the county level, administrative regions can vary widely in size, which can influence the visual perception of count data, such as the change in population counts between two years. With choropleth maps, this often translates to big areas, even those smaller percentages of change over the two years, getting undue emphasis by being grouped into the higher, or darker-colored, classes. To offset this potential problem, cartographers do not map count data using this method; instead differences in the areal extents of regions are processed out statistically prior to visualization by changing the count data to derived data.

To standardize our change in population data, we will convert it to a percent change over the ten years.

1. Right-click *Census_South_2000_2010* and choose *Open Attribute Table*
2. To calculate the percent change in population between 2000 and 2010, right-click the field *PPOP_CHG* and choose *Field Calculator*

3. Click *Yes* to dismiss the warning dialog box
4. In the bottom window labeled PPOP_CHG =, type ([**POP_CHG**]/[**POP2000**]) * **100** and click *OK*
5. Close the attribute table

QUESTION 6.2

Which of the following data sets are NOT standardized and, therefore, are data sets that should NOT be mapped using the choropleth map technique?

a. Percentage of population that is Caucasian by state
b. Number of people per square mile by state
c. Number of restaurants by state
d. Number of crimes per 100,000 people by state

CLASSIFY YOUR ATTRIBUTE DATA

Data visualized using the choropleth map technique are typically classified by grouping the data values into a smaller set of four to six classes. This generalizes and highlights the major patterns or trends in the data for the map user. It also limits the number of color values the user has to discriminate between reliably. Choosing a classification method and an appropriate number of classes relies on the cartographer's knowledge of those methods and how they relate to the data's statistical distribution.

1. Right-click the data layer in the *TOC* and choose *Properties,* then select the *Symbology* tab
2. To the left in the dialog box, click on *Quantities – Graduated Colors*
3. Under the *Fields* area, change *Value* to PPOP_CHG
4. Under the *Classification* area, click *Classify*
5. Modify the number of classes and the classification method until you have a combination you believe tells the story of population change in the south most reliably
6. Click *OK*

QUESTION 6.3

Statistics for the field are computed in the *Classification Statistics* area. Record these below:

a. Count: _____
b. Minimum data value: _____
c. Maximum data value: _____
d. Mean: _____
e. Median: _____

This histogram is a graphic display of the distribution of the attribute data, where the data values are plotted along the x-axis and the frequency with which they occur is plotted along the y-axis. There are several common distribution patterns (Figure 6.2); and they can be described in terms of their shape, center, and spread among other features. The shape of the histogram can help you choose a reliable classification method. Skewed distributions, for example, often make use of *Natural Breaks* to classify the dataset. Multipeaked distributions also make use of Natural Breaks. *Equal Intervals* and *Quantiles*, on the other hand, which do not take the shape of the distribution into account when grouping data values, tend to serve uniform and normal distributions more effectively.

FIGURE 6.2
Source: Elisabeth Nelson

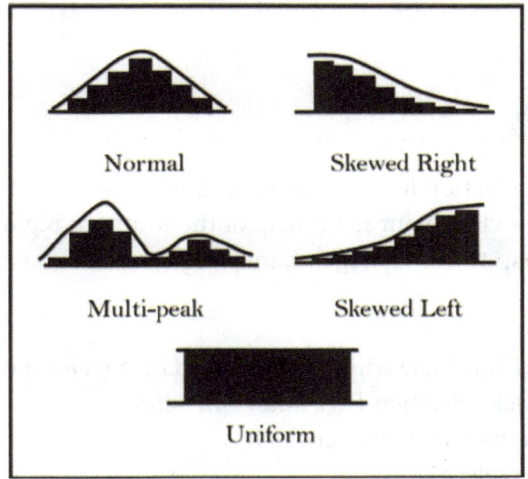

QUESTION 6.4

The histogram for this dataset has this shape:

a. More uniform in appearance
b. More normal in appearance
c. More skewed to the left (tail points to the left) in appearance

SELECT A COLOR SCHEME

The next step is to select a color scheme that will shade your polygons according to the data class to which they are assigned. In choropleth mapping, when ordered or numerical data are being mapped, the rule is to pick a scheme in which a light-to-dark ramp of shades predominates. Lighter shades should be assigned to the lower data values, while darker shades should be reserved for higher data values.

1. Under *Color Ramps*, choose an appropriate color scheme
2. Click *OK*

QUESTION 6.5

When the goal is to showcase variation in numerical data, like changes in population, the best choice is the color scheme that:

a. emphasizes a progression of light to dark values that match the progression of lower to higher changes in the numerical data
b. emphasizes different color hues to differentiate the different classes of numerical data
c. emphasizes a progression of dark to light values that match the progression of lower to higher changes in the numerical data

Design a Map Overlay

Right now, you have a data visualization in the *Data View*, but you will need a printed final product for your employer. This product should include your main map of the southern region as defined by the U.S. Census Bureau. You will also want a map title, map legend, map scale, source statement, and credits statement for your finished design.

1. In the bottom left of the *Data View* window, click the *Layout View* button
2. From the Main Menu, click *File*, then *Page and Print Setup*
3. Change Orientation: to *Landscape*, then click *OK*
4. Click once on the map to select it
5. Click and drag the map to the middle of the newly oriented page
6. Click on the middle blue handle at the top of the map and drag it downward until the frame is on the Page
7. Repeat with the middle blue handle at the bottom of the map
8. Right-click the map and choose *Properties*
9. Under *Frame*, change the color of the border to *No Color* and click *OK*

ADD A NEATLINE

1. From the Main Menu, click *Insert* and choose *Neatline*
2. Click *Place Inside Margins*
3. Click *OK*
4. Right-click the selected rectangle that results and choose *Properties*, then *Size and Position*

While the neatline can cover the entire page, it is often a better idea to work within a smaller extent – for one, sometimes printers won't print all the way out to the edges of the paper. In this case, your employers want a final map that is 9 x 7 inches.

5. Make the *width* = 9 and the *height* = 7 and click *OK*
6. Click and drag the neatline to the center of the page
7. Use the *Page Zoom* tool to zoom in on it

RESIZE THE MAIN MAP

1. Select the *Arrow* tool, and then select the map by clicking on it
2. If needed, drag the map until it is in the center of the neatline
3. Use the *Data Zoom In* tool to increase the scale of the map to around **1:20,000,000**
4. Resize the frame of the map box as needed to see the entire map – you can do this by clicking and pulling on the blue tabs that show when the map is selected

ADD A MAP TITLE

1. From the Main Menu, click *Insert* and choose *Text*–a small text box will be created and placed in the center of the map
2. Use the *Page Zoom tool* to zoom in and see the text
3. Use the *Select Elements tool* to reposition the text box as needed
4. Double-click the text box to modify the text–you can change the font, the style, the size, even whether you want it to be single or multiple line
5. Change the text to *Population Change in the Southern Census Region* and click *OK*
6. Double-click the text on the page to edit it
7. Click *Change Symbol* to change the typeface, typesize and typestyle to suit your layout, then click *OK*
8. Repeat the process to add a second line to the title that says: *United States, 2000-2010*
9. Position both lines of text on the map where you think you want them to be
10. Hold the *Shift* key down and click both lines of text to select them, then right-click and select *Align –Center, Left, or Right* to line up the individual text pieces as you wish
11. From the *File* menu, choose *Save* to save your map document

ADD THE MAP LEGEND

1. In the Table of Contents, right-click the *Census_South_2000_2010* layer and choose *Properties-Symbology*
2. In the *Symbol* box, click the *Label* heading, then *Format Labels* (Figure 6.3)

FIGURE 6.3

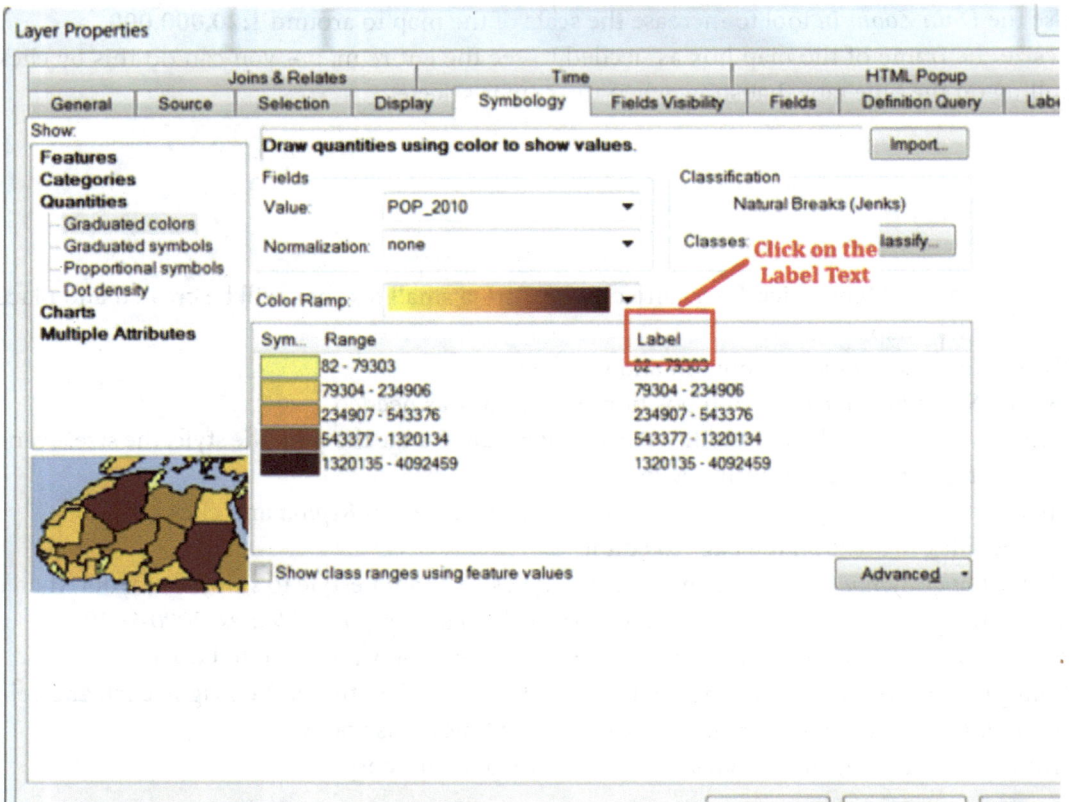

3. For *Number of Decimal Places,* choose **1** and click *OK* then click *OK* again
4. From the Main Menu, click *Insert* and choose *Legend*
5. In the Legend Wizard, verify the *Census_South_2000_2010* layer is in the *Legend Items* Column
6. Click *Next*
7. For the *Legend Title* enter *Percent Population Change by County*
8. Make the font match that of the title; change the size as needed
9. Click *Next,* then *Next* again
10. Change the *Patch Width* to 15 and *Patch Height* to 9 and click *Next*
11. Change *Patches (vertically)* to 3
12. Click *Finish*
13. Drag the legend to the general area where you think you might like it–don't let the size/shape constrain you, you can change that later on
14. Right-click the legend and choose *Convert to Graphics*
15. Right-click the legend again and choose *Ungroup*
16. Select and delete the unneeded text lines
17. Re-position the legend title and legend items as needed–you can change the size and number of lines the legend title takes up, you can also change the order of the legend boxes if you like. For example, you might want to change to a horizontal layout or perhaps a multi-column layout.

18. Select your legend objects, right-click and choose *Group* to group them back and move them as a single object

ADD A SCALE BAR

1. In the *Page Layout* view, click to select the main map of the southeast U.S.
2. From the *Insert* menu, choose *Scale Bar*
3. Select a style, then click *Properties*
4. Under the *Format* tab, change the font size to **7**
5. Under *Scales & Units,* change the *number of divisions* to 3 and the *number of subdivisions* to 0
6. Under *Numbers and Marks*, set the *Numbers Frequency* to *divisions*
7. Click *OK,* then *OK* again
8. Use the blue tabs to resize the bar so that it takes up approximately 1 ½ inches in length
9. Position the element where you think you would like it to be

ADD SOURCE AND CREDIT STATEMENTS

1. From the Main Menu, click *Insert* and choose *Text*
2. Use the *Page Zoom tool* to zoom in and see the text
3. Use the *Select Elements* tool to reposition the text box as needed
4. Double-click the text box to modify the text–you can change the font, the style, the size, even whether you want it to be single or multiple line
5. For the Source Statement, enter: *Source: U.S. Census Bureau, 2010* and click *OK*
6. Repeat the process to add a Credits statement that says: *Credits: Your Name, Current Year*
7. Position both lines of text on the map where you think you want them to be
8. Use the *Select Elements* tool to select and reposition all the map elements as needed

CHECK THE VISUAL BALANCE AND HIEARCHY OF YOUR MAP

1. Use the *Zoom Whole Page* tool on the Layout Toolbar to zoom out so your entire map shows within the Layout View
2. From the *File* menu, choose *Save*
3. Now, from the *File* menu, choose *Export Map*
4. In the *Save In* drop-down box at the top of the dialog, navigate to the same location that you saved the original map document
5. Under the *Save as Type* drop-down box, choose *PDF*
6. Name your file to match the name of the original map document, then click *Save*
7. Double-click the *Computer* icon your desktop
8. Navigate to the location where you think you saved your file and verify that it is there

Note: To receive full credit during the grading process, the map you export should reflect the latest changes as directed by the exercise.

Evaluating Your Map

    a.   Did you choose an appropriate equal-area projection?

    b.   Did you re-center the map appropriately?

    c.   Did you standardize your data correctly?

    d.   Did you classify your data appropriately?

    e.   Did you choose an appropriate color scheme to highlight your data classes?

    f.   Did you add an appropriately designed and placed map title?

    g.   Did you add an appropriately designed and placed map legend?

    h.   Did you add an appropriately designed and placed scale bar?

    i.   Did you add an appropriately designed and placed source statement?

    j.   Did you add an appropriately designed and placed credits statement?

    k.   Are all elements located inside the neatline?

Chapter 7

Spatial Data in the Digital Environment

Over the last few decades, computer technology has revolutionized mapmaking. Now, both reference maps, the topographic quadrangle and the thematic map, have not only a printed presence but also a digital presence. Moving from a real-world, printed map to a digital representation is known as *data modeling*. Data modeling for spatial data got its official start with the development of the *Canada Geographic Information System (CGIS)*. CGIS was the first industry-scale computer-based GIS, and most GIScientists recognize it as a primary milestone in the development of today's GIS packages. The concept of CGIS was born in 1963, when the Canadian government became concerned with decision-making practices regarding land use in its country. Maps with sufficient detail for making land use decisions were accessible, but the manual process of analyzing and interpreting them to arrive at these decisions was too time-consuming and labor-intensive for the government to support the tasks. Even if the money had been available, in reality there weren't enough people trained in these skillsets to perform the work for such a large expanse of land.

The idea of harnessing computer power to conduct these analyses originated with *Roger Tomlinson*, who had encountered similar issues while working on projects for an aerial survey company. As someone who saw this need as critical, he began experimenting with the notion of converting a printed map into a format that a computer could analyze. Computers can't "see" graphics and images like we can; their reality is a world of numbers. One of the first challenges Tomlinson faced, then, was how to create a map *model* the computer could "see" and interpret that would mimic what we as humans see and interpret when the same map is put in front of us.

Data Models

Data modeling consists of a series of steps; Worboys (1995) has one of the more easily understood descriptions (Figure 7.1). He breaks the process down using four hierarchical levels, which range from *Application Domain* (the original printed map) to the *Physical Computer Model*: (actual computer code that implements the model). The two levels of most interest here are the second and third levels. The second level, *Conceptual Computer Model*, translates a map into a numerical model. Two basic models that have evolved over the years are the *raster* data model and the *vector* data model. The third level, Logical Computer Model, translates the conceptual model into a computer-based data structure.

Raster Data Model. Tomlinson's first stab at designing a conceptual computer model for CGIS was based on scanning the needed data. He had the area features traced from the original maps to create an uncluttered overlay, and then used a drum scanner to scan the new line work into a digital format. During scanning, the scanner head moved back and forth across a revolving drum onto which the overlay was attached, detecting reflected light intensity levels every 1/250 of an inch. Each sampled light intensity level was stored as a *pixel*, or picture element, using a grid format. For each cell in the grid, the computer recorded either a value of 1 (presence of a line – no light reflected due to ink presence) or a value of 0 (absence of a line – light reflected due to absence of ink). These values, in the order in which they were written, produced a grid of cells representing the original line work. Each cell (and corresponding value) in the grid had an assigned row and column number that gave it a unique location in the file. The result was a prototype for the *raster data model* (Figure 7.2). A more generic description would be to describe the process as overlaying a grid onto the printed map, then numerically coding the primary geographic feature occupying each cell.

FIGURE 7.1
Worboy's (1995) data modeling hierarchy.
Source: Elisabeth Nelson

Application Domain

↓

Conceptual Computer Model

↓

Logical Computer Model

↓

Physical Computer Model

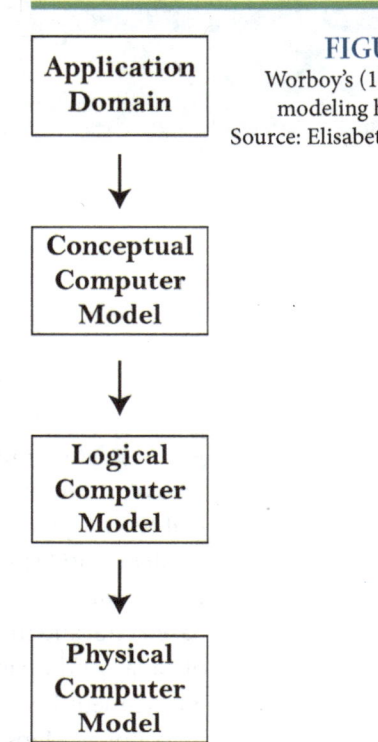

FIGURE 7.2
Raster Data Model.
U.S. Geological Survey,
Department of the Interior.
Source: Elisabeth Nelson

The first raster data model was a success from the standpoint of converting geographic features to numbers, but beyond simply spitting out what Tomlinson already had in printed form, it was severely limited. There was no way, for example, to record any descriptive data about the scanned features, and Tomlinson knew he would need this data to calculate statistics and produce reports about the natural resources of Canada. He also envisioned a model that would 1) convert row and column numbers to latitude and longitude to tie features to real world locations, 2) store descriptive data about each feature, 3) calculate areas, 4) overlay and/or merge features, 5) select feature subsets, and 6) create buffers around features.

Vector Data Model. Tackling the descriptive data issue first, Tomlinson decided to keep the locational information about the areas separate from the descriptive data about them. To link the descriptive data to the feature, he needed to treat each feature as a separate entity rather than a series of grid cells. His solution was to use a geometric shape, the polygon, to represent the different map features (Figure 7.3). To convert the scanned features to geometrical figures based on digital coordinates, he placed the traced overlay of area features, along with the original map, onto a *digitizing table* (an electronically-sensitive device), and used the computer to record the latitude and longitude of the four corner points of the original map, along with a point inside of each area feature. He then let the computer figure out the real-world coordinates of each of the scanned features. The points inside each of the resulting polygons were assigned *feature identification numbers (FID)*, which were also embedded in the tables used to store the descriptive, or *attribute*, data associated with each feature. This model was the prototype for the *vector data model*, a model in which graphics are defined using geometric primitives instead of grid cells.

FIGURE 7.3
Vector data model.
U.S. Geological Survey,
Department of the
Interior.
Source: Elisabeth Nelson

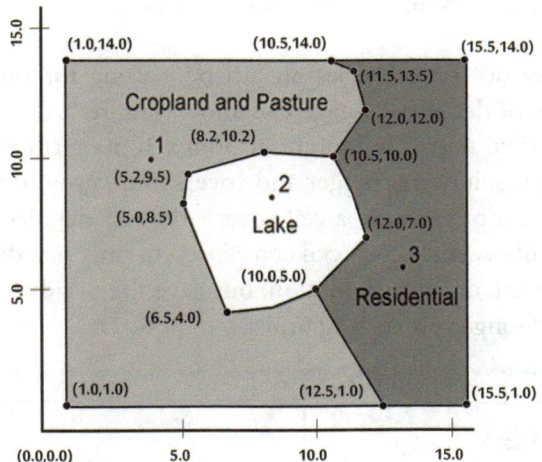

Data Structures

Both the raster and vector data models are concepts until they are implemented by creating more specific *data structures* that can be used inside the GIS. A variety of structures are available for implementing both raster and vector data models. These structures, which can affect both the efficient processing of data and the amount of data storage required, are members of Worboy's *Logical Computer Models*, the third level of his hierarchy.

Raster Data Structures. Raster data structures are implemented using *arrays*, which are groups of contiguous cells, each holding a value of the same data type. These values are accessed using indices, or row and column addresses, that represent location coordinates. Tomlinson's prototype architecture, known as a *bitmap image*, represents the simplest structure for raster data. The grid cell values of this format are numerical, but are limited to ones (white) and zeroes (black) to indicate the presence or absence of a particular feature. In 1967, researchers at the *Harvard Lab for Computer Graphics and Spatial Analysis* modified this prototype to represent and analyze digital landscapes for environmental planning applications. While conducting a regional study, they found that the land use and elevation data available for analysis was stored in a grid cell format, leading them to expand on Tomlinson's structure to take advantage of this (Chrisman, 2006). They developed a grid-based input system that could record both *discrete* data values, such as multiple land use codes, and *continuous* data values like elevation measurements. In these cases, the data values were expanded from the simpler 0/1 structure to include a variety of integers and floating-point data values to represent changes in discrete and continuous data, respectively.

For each cell in the grid to house these expanded data values requires manipulating each pixel's *bit depth*. Bit depth refers to how much computer memory (or how many unique data values) a pixel in a raster file can store. It is based on the equation 2^n, so a bitmap with a bit depth of 1, like Tomlinson's raster prototype, has pixels with enough memory to store 2^1, or 2, unique data values. Files with a bit depth of 2, then, can store 2^2, or four unique data values. Some of our current raster files, like Esri's Grid format, can store 2^{32}, or 4,294,967,296 unique data values per pixel.

Each cell in the array not only houses an attribute value for the area, its size is correlated to the level of detail the image can show (Figure 7.4). This characteristic is called *spatial resolution*. A grid cell might be 30m x 30m or 1m x 1m, for example. The cell with the 30m resolution is bigger and covers more ground than the cell with the 1m resolution. The more ground a cell covers, the less detailed the features and patterns of the area will be, since each cell can represent only one data value. Smaller cells provide better detail, or finer resolution, but have the disadvantage of requiring more data storage and longer processing times.

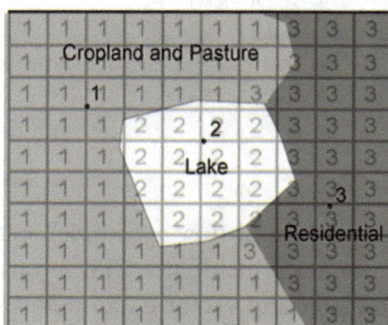

Smaller Cell, Finer Spatial Resolution

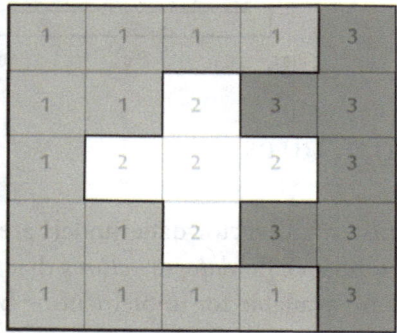

Larger Cell, Coarser Spatial Resolution

FIGURE 7.4
Comparing spatial resolutions and associated detail changes. *U.S. Geological Survey, Department of the Interior.* Source: Elisabeth Nelson

EXPLORING GEOSPATIAL TECHNOLOGY

Vector Data Structures. In Tomlinson's vector data structure, the polygon was the geometric figure used to describe areal features mathematically. Like his raster prototype, this one was also modified by researchers at the *Harvard Lab for Computer Graphics and Spatial Analysis* in the 1960s. In a reaction to an attempt to produce the first graphic output for a mapping system, researchers designed a new mapping package known as SYMAP. This package, which produced thematic maps and output them to a line printer, used a vector data structure that supported line and point features as well as polygons. Attribute values, like those in the CGIS prototype, were kept in a separate file from the geography with the two files linked by a feature ID (Chrisman, 2006).

The points, lines, and polygons comprising this data structure are represented by (x,y) coordinates that are tied to a spatial reference system such as latitude and longitude (Figure 7.5). The point, represented by a single (x,y) coordinate, is the simplest geographic feature. It could be a city, or an archaeological site, perhaps. Lines, such as roads or railroads, are represented by storing strings of coordinates that must be connected by the computer when drawn. Areas, or polygons, are also represented by storing a string of coordinates, except in this case the first and last coordinate pairs are the same, resulting in a closed figure when drawn. Storing these features using coordinates allows us to 1) transfer the data to the computer, 2) use the data to redraw the map digitally, and 3) perform basic calculations, such as the distance between two features.

FIGURE 7.5
Non-topological vector data structure.
Source: Elisabeth Nelson

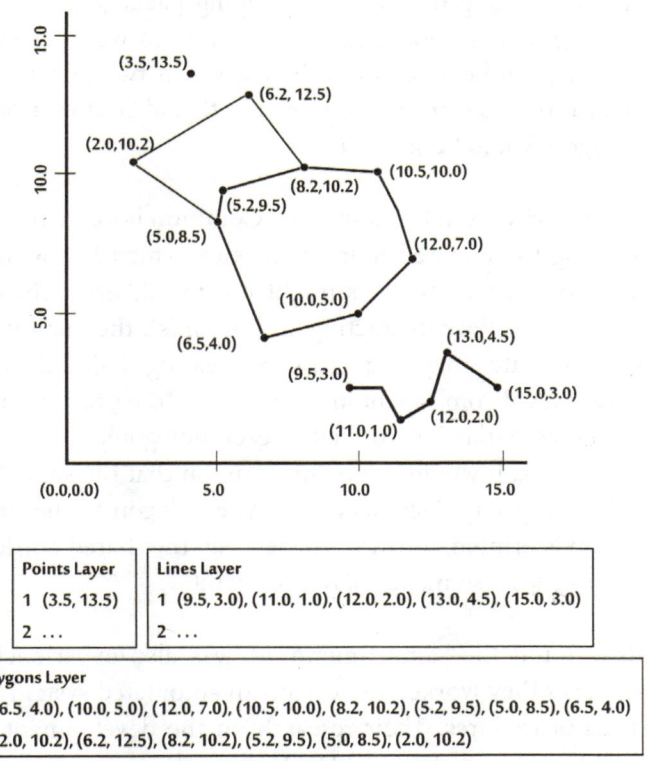

Points Layer	Lines Layer
1 (3.5, 13.5)	1 (9.5, 3.0), (11.0, 1.0), (12.0, 2.0), (13.0, 4.5), (15.0, 3.0)
2 ...	2 ...

Polygons Layer
1 (6.5, 4.0), (10.0, 5.0), (12.0, 7.0), (10.5, 10.0), (8.2, 10.2), (5.2, 9.5), (5.0, 8.5), (6.5, 4.0)
2 (2.0, 10.2), (6.2, 12.5), (8.2, 10.2), (5.2, 9.5), (5.0, 8.5), (2.0, 10.2)

This next generation the vector data structure was aptly named *"cartographic spaghetti"*, after a comment made by Nick Chrisman in which he described the data structure adopted by the CIA for its World Data Bank as "...unstructured as a plate of spaghetti" (Chrisman 2006:105). Esri's *shapefile* format is a modernized version of this vector data structure. In the shapefile, each set of features on a map (points, lines, polygons) are stored in their own data file using these coordinates. Stored in this manner, the computer can draw the features, but it has no way to differentiate them symbolically within any data file. For this to occur, we need to supply the computer with *attribute data* for each feature. Attribute data are recorded in a flat file, or spreadsheet form, for each geographic feature. This file is linked to the file of feature geometry using a Feature ID Number (FID). The FID is included in both the geometry and attribute files, and is a unique code supplied to each geographic feature (Figure 7.6). Because it exists in both files, the computer can use it to associate the correct attribute data with the correct geometry. For example, by linking city coordinates (points) to a table that lists the population for each city, we can use different circle sizes for each city, depending on how many people reside there. This setup, where we use multiple tables that are linked by FIDs, is called a *georelational database*.

Cartographic spaghetti is considered a simplistic vector data structure by today's standards. Its initial purpose was to provide the functions necessary to reproduce a map using the computer; functions to analyze spatial relationships

Points Layer		Points Attributes		
FID	Coordinates	FID	Name	Population
1	(3.5, 13.5)	1	Colfax	2,469
2	...	2	...	

FIGURE 7.6
Example of a georelational database setup.
Source: Elisabeth Nelson

were missing entirely, making this more a mapping package and less a GIS. Even as a mapping package, it was soon clear that there were weaknesses in its design. Polygons, for example, were entities unto themselves; if two polygons happened to share a border, such as the state border between North and South Carolina, the shared border of each polygon would be stored twice.

At least two problems arise from this situation. Common borders between polygons, for instance, are going to be stored more than once, which is a waste of computer memory. Further, those shared borders might not match precisely, creating *slivers* and *gaps* upon printing that are distracting and diminish the quality of the map. In the earlier days of computer mapping and GIS, creating digital data to be used by these systems was a critical component in the process. To reproduce printed maps as numerical structures, someone has to *digitize* each polygon's boundaries. A tedious and time-consuming task, it would not be uncommon that the shared border would differ at least slightly in point placement from one polygon to the next. When that happens and the map is printed, mismatches between the shared borders are likely to be visible. Slivers and gaps also make analysis difficult.

The U.S. Census, around the same timeframe, was also experiencing redundant digitizing operations as they worked to develop an automated system to increase the efficiency of census procedures. Their research on the development of geographic base files eventually produced the GBF-DIME file, which incorporated the system of *map topology* to eliminate redundant digitizing and enhance analytical and map production procedures (U.S. Census, 2013). GBF-DIME stands for *Geographic Base File – Dual Independent Map Encoding* and map topology refers to the mathematical concept that " . . .explicitly expresses the spatial relationships between features, such as two lines meeting perfectly at a point and a directed line having an explicit left and right side" (Chang, 2008:7).

Topology is a critical concept for GIS; building it into base files as they are converted from print to digital formats allows digitizers to eliminate shared boundaries when building polygons and helps them reduce graphic clutter. With the DIME encoding scheme, intersections were treated as points, streets as lines, and blocks or areas as polygons. Intersections, or *nodes*, and areas were each numbered, then street segments were tagged with the areas to the left and right of each segment, as well as address ranges for both sides of the street, and for the two nodes (*From-nodes* and *To-nodes*) each segment connected (Figure 7.7). The street segment, also known as a *chain* or *arc*, is the basis for this system; the tagged areas explicitly encode spatial *adjacency* relationships and the From and To nodes explicitly encode *connectivity* and *contiguity* relationships. During the 1970s, GBF-DIME files were created for all cities in the U.S. This is the structure upon which many of the modern vector systems are built, including the Census' *Topologically Integrated Geographic Encoding and Referencing* (TIGER) files, the USGS's *Digital Line Graphs* (DLGs) and ESRI's *coverages*.

FIGURE 7.7
Topological data structure.
Source: Elisabeth Nelson

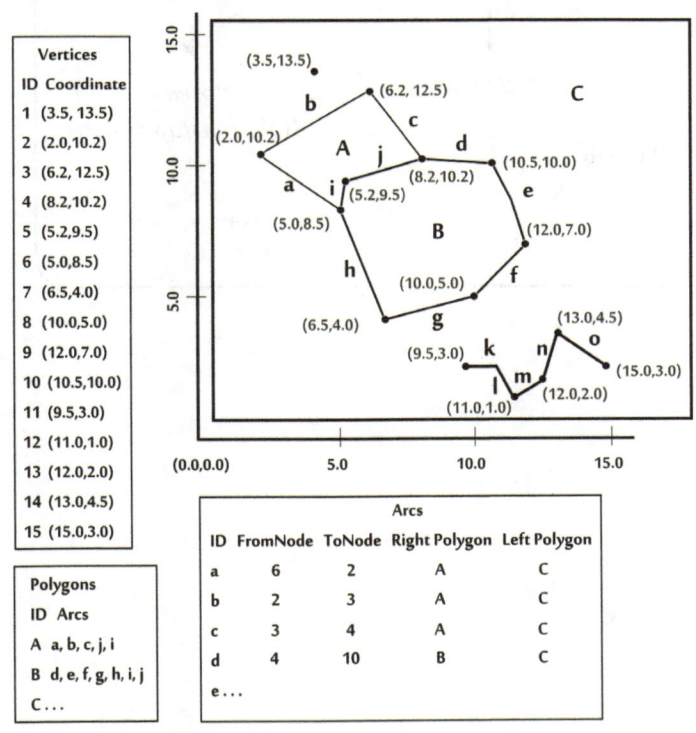

Vertices	
ID	Coordinate
1	(3.5, 13.5)
2	(2.0,10.2)
3	(6.2, 12.5)
4	(8.2,10.2)
5	(5.2,9.5)
6	(5.0,8.5)
7	(6.5,4.0)
8	(10.0,5.0)
9	(12.0,7.0)
10	(10.5,10.0)
11	(9.5,3.0)
12	(11.0,1.0)
13	(12.0,2.0)
14	(13.0,4.5)
15	(15.0,3.0)

Polygons	
ID	Arcs
A	a, b, c, j, i
B	d, e, f, g, h, i, j
C	...

Arcs				
ID	FromNode	ToNode	Right Polygon	Left Polygon
a	6	2	A	C
b	2	3	A	C
c	3	4	A	C
d	4	10	B	C
e	...			

The newest vector data structures are object-oriented implementations, such as ArcGIS' *geodatabase* (Figure 7.8). A geodatabase, at its most functional level, can be thought of as a container that houses data, both spatial and attribute together, in the context of a georelational database. Non-spatial data are stored in *geodatabase tables* and spatial data in *feature classes*. Each feature class may be comprised of only one spatial data type (points, lines, or areas), and whatever that type is, each feature in the class must have the same attribute fields associated with it. Feature classes that are tied to each other via spatial relationships, such as a county feature class and a corresponding census tract class, can be stored together as a *feature dataset* to make modeling those relationships in the GIS easier using *geodatabase topology* rules. Census tracts, for example, make up a county, so they share some boundaries with the county feature class. If the boundary of a tract changed, that change might also need to be made at the county level as well. By storing both in a feature dataset, where a common spatial reference system is required, these updates are easier to implement (Esri, 2013). *Relationship classes* can be used to link feature classes and geodatabase tables; an example here might be the need to create a permanent association between a feature class of parcels and a non-spatial table of parcel owners.

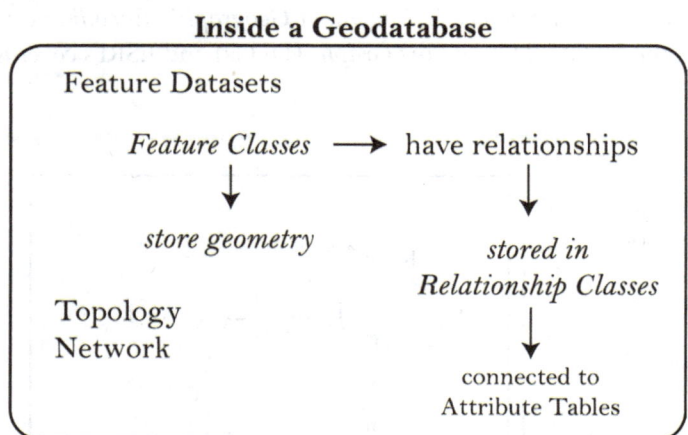

Inside a Geodatabase

Feature Datasets

Feature Classes ⟶ have relationships

↓

store geometry

Topology
Network

stored in
Relationship Classes

↓

connected to
Attribute Tables

FIGURE 7.8
Basic view of the geodatabase.
Source: Elisabeth Nelson

Additional Readings

Chang, K. 2008. *Introduction to Geographic Information Systems*, 5th ed. McGraw Hill Companies, Inc: New York, New York.

Chrisman, N. 2006. *Charting the Unknown: How Computer Mapping at Harvard Became GIS.* Esri Press: Redlands, California.

Cote, P. *GIS Manual: GIS Tutorials and Resources.* Last accessed June 18, 2013. http://www.gsd.harvard.edu/gis/manual/data_structures/

Davis, B. 2001. *GIS: A Visual Approach*, 2nd ed. Onword Press: Albany, New York.

DeMers, M. N. 2000. *Geographic Information Systems,* 2nd ed. John Wiley & Sons, Inc: New York, New York.

Environment Canada. 1973. The Canada Geographic Information System. *Cartographica*, v. 10(3): 62-86.

Esri. *ArcGIS Desktop Help.* Last accessed June 19, 2013. http://resources.arcgis.com/en/help/main/10.1/

Esri. *Company History.* Last accessed March 17, 2009. http://www.esri.com/company/about/history.html

Foresman, T.W. (ed.) 1998. *The History of Geographic Information Systems: Perspectives from the Pioneers.* Prentice Hall: Upper Saddle River, New Jersey.

National Academy of Sciences. 2006. *Beyond Mapping: Meeting National Needs Through Enhanced Geographic Information Science.* National Academies Press: Washington D.C.

NCGIA. *The GIS History Project.* Last accessed June 19, 2013. http://www.ncgia.buffalo.edu/gishist/DIME.html

Shashi, S., and H. Xiong. *Encyclopedia of GIS 2008.* Last accessed June 18, 2013. http://link.springer.com/referenceworkentry/10.1007%2F978-0-387-35973-1_1080/fulltext.html

Slocum, T., McMaster, R., Kessler, F., and H. Howard. 2008. *Thematic Cartography and Geovisualization,* 3rd ed. Prentice Hall: Upper Saddle River, New Jersey.

Wise, S. 2002. *GIS Basics.* Taylor & Francis, Inc.: New York, New York.

Worboys, M. F. 1995. *GIS: A Computing Perspective.* Taylor & Francis: London, England.

Exercise 7
Spatial Data, Digital Environment

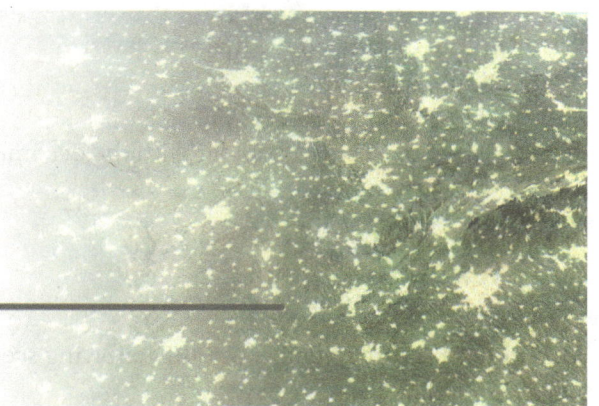

The purpose of Exercise 7 is to explore the different data structures associated with GIS. You will work with both raster and vector data structures to understand the different ways in which spatial data can be represented in a computer.

OBJECTIVES

- Examine the different structures of vector data
- Examine the different structures of raster data
- Learn how to use metadata effectively

SOFTWARE INFORMATION

Introducing ArcMap uses Esri's **ArcGIS 10.6** software. **ArcGIS** is a commercial GIS package, available in geospatial labs affiliated with schools that have a campus-wide site license for the software. Instructors at these campuses may also request 1-year student versions of the software at http://www.esri.com/landing-pages/education-promo.

DATA

The data for the exercises is available from the Kendall Hunt Student Ancillary site. See the inside front cover for access information. You may also be directed to download the data from a different location by your instructor.

1. Download the data as instructed by your instructor
2. Save the file to a location where you have read/write privileges (USB key, home computer, class server)

The file you just saved is in a compressed (*.zip) format, and was created using **7-zip** freeware (http://www.7- zip.org). To use the data for this exercise, you must decompress it using the same software.

3. Locate the zipped file that you saved
4. Right-click the file
5. Choose *7-Zip - Extract Files*
6. Click *OK* to create a folder with the decompressed data

Scenario Summary

The exercise is contextualized by the following scenario:

Fracking is coming to an area near you. More properly termed *hydraulic fracturing*, fracking is a method of extracting natural gas from an organic-rich, fine-grained sedimentary rock known as shale. This method involves injection chemicals, sand, and millions of gallons of water under high pressure directly into the ground to release natural gas in the shale deposits.

Detractors claim that the negative effects of fracking outweigh any possible benefits. Some the effects they list are that:

- The mixture of chemicals, sediment, and natural gas injected may make its way to the surface and enter rivers
- Byproducts and natural gas can leach into groundwater
- Fracking can weaken existing fractures in the rock, known as faults, leading to an increase in earthquakes

The North Carolina State Legislature has passed a new law allowing fracking. The town of Oak Ridge, located in the northwest quant of Guilford Count in the central part of N.C., wants to be proactive and has asked you to examine the data they have collected that might be used for future mitigation and emergency management strategies. This exercise examines the data structures and metadata associate with geospatial datasets that could be used in a fracking analysis.

Metadata

GIS datasets consist not only of spatial data and attribute data, but also metadata. Metadata is typically included as a text file and describes relevant details about the spatial data to which it is attached. These details can be crucial to evaluation the data and deciding whether to use it in the context of specific GIS project. In ArcGIS, metadata is most often accessed using ArcCatalog, a sister application to ArcMap that is used to manage spatial data. Similar to Microsoft's Windows Explorer, ArcCatalog organizes your spatial data using a tree view and provides detailed information about datasets.

1. Start ArcCatalog by double-clicking the icon on your desktop or by clicking Programs > ArcGIS > ArcCatalog. **Be sure to start ArcCatalog outside of an ArcMap session.**
2. In the catalog tree on the left side of the display, right-click *Folder Connections* and choose *Connect to Folder*

3. Navigate to the location where you saved the *Chapter07* folder of the data for this exercise
4. Click once to select the *Chapter07* folder, then click *OK* to create a folder connection that will allow you to access the data inside the folder
5. Click to select the folder in the *Catalog Tree*
6. Click the *Contents* tab to the right of the catalog tree to view the *Contents Panel*

QUESTION 7.1

Which of the following files is NOT listed in the Contents Panel once you have click on the *Chapter07* folder?

a. NorthCarolina.gdb
b. Temp (a Folder)
c. nc_trcts
d. OakRidge_Subset.tif

The toolbar underneath the menu in ArcCatalog contains four buttons that control how the Contents Panel is viewed (Figure 7.1):

FIGURE 7.1

1. Click on each of them to see how the view changes
2. Click the *Preview* tab at the top of the display window in *ArcCatalog*

Some types of files are not geospatial in nature and cannot be previewed in ArcCatalog. These include Windows folders, text documents, and spreadsheets.

1. Click the + sign to the left of the *Chapter07* folder
2. Click the + sign next to *NorthCarolina.gdb*
3. Click nurhm (a shapefile denoted by the .shp extension) to select it, then click the *Preview* tab to view its contents
4. Under the *Preview* drop-down menu at the bottom of the display window, click and choose *Table*

QUESTION 7.2

Scroll across the columns of attribute and find FAC_TYPE. FAC_TYPE records what type of facility is mapped at each point in the dataset.

The facilities mapped in this dataset are all related to homes for the elderly and nursing facilities.

    a.   True
    b.   False

EXAMINING THE METADATA

1. Under the *Customize* menu, choose *ArcCatalog Options*
2. Select the *Metadata* tab
3. Under *Metadata* tab, make sure that **North American Profile of ISO19115 2003** is selected as the *Item Description*
4. Click *OK*
5. Click the *Description* tab to access the contents of the *Description Panel* for the *nurhm* shapefile located in *NorthCarolina.gdb* geodatabase
6. Under the heading *ArcGIS Metadata*, there are several subheadings – scroll to examine these and answer the following question

QUESTION 7.3

Under which subheading will you find the projection information for this file?

    a.   Resource Details
    b.   Spatial Reference
    c.   Spatial Data Properties
    d.   Extents

In many cases, rather than populating all the metadata fields in the *Description Panel*, the originators of the data will include their additional information in the form of a text document that you can download with your data.

1. In the *Catalog Tree*, click the + sign to expand the *Temp* folder
1. Expand the *Docs* folder
1. Double-click nurhm.txt in the *Catalog Tree* to open the file

QUESTION 7.4

What is the title of this dataset?

a. Assisted Living Facilities
b. Nursing Homes
c. Hospitals
d. Adult Day Care Facilities

QUESTION 7.5

If you scroll down, you will find information for each of the attributes in the table. Use the *Find* command under the *Edit* menu to search each instance of FIPS in the document.

Based on your search, the FIPS attribute definition identifies the FIPS as a 2-digit code that tells you in which city the point exists.

a. True
b. False

Vector Data

A data model is a conceptual representation of reality. When that model is implemented in a computer, it is called a data structure. In ArcGIS, there are three vector data structures that are popular associated with the shapefile, the coverage, and the geodatabase. Each of these is a view of reality that uses points, lines, and polygons to represent spatial features in the environment.

SHAPEFILES

1. Click *snha* (a shapefile; *.shp) and select the *Description tab*
2. Scroll through the information

QUESTION 7.6

What are the Geographic Coordinate Reference System and Projection?

a. GCS_WGS_2984; WGS_1984_UTM_Zone_12N
b. CGS_North_America_1983; NAD_1983_Contiguous_USA_Albers
c. CGS_North_America_1983; NAD_1983_StatePlane_North_Carolina_FIPS_3200
d. GCS_Sphere; Sphere_Mercator

QUESTION 7.7

Under *Spatial Data Properties – ArcGIS Feature Class Properties*, identify the following: What is the geometry type?

a. Point
b. Line
c. Polygon
d. TIN

QUESTION 7.8

This spatial dataset has topology embedded in it.

a. True
b. False

QUESTION 7.9

In the text document associated with this data layer, find the *Abstract*. What is the data layer displaying?

a. Significant Natural Heritage Areas
b. Substantial National Hardwood Areas
c. Significant National Historic Areas
d. Substantial Natural Honey Areas

COVERAGE

A coverage is made up of a set of files, where each file contains information specific to a collection of common features, also known as a *feature class*. More than one feature class is common in a coverage, as many features, such as those comprised of polygons, require both a polygon and a line feature class to define the feature.

gc_parcel_cv2 is a coverage in your data folder. A parcel is defined as a tract of land; this coverage displays some parcels for Guilford County, NC. A county in the piedmont or central portion of NC that just happens to include the town of Oak Ridge.

QUESTION 7.10

Click the + sign next to *gc_parcel_cv2* in the TOC. Which the following is NOT a feature class in this coverage?

a. Arc
b. Label
c. Polygon
d. Isoline

1. Click to select the *arc* file
2. Under the *Preview* tab, examine the data in the *Table* view

QUESTION 7.11

By examining the attributes in the *arc* file, you can indirectly determine whether a set of files has topology embedded in the dataset's geometry.

gc_parcel_cv2 has topology embedded in its geometry.

a. True
b. False

GEODATABASES

Your data folder *also* includes a geodatabase called *NorthCarolina.gdb*; it represents the most current vector data structure used by ArcGIS. Geodatabases can house feature classes, raster datasets, and feature datasets.

Feature datasets are collections of feature classes that are related whether thematically or spatially, and they must share the same coordinate system. They are also the containers in which you can create topological rules and geometric networks, among other functions, within and between feature classes.

1. Click the + sign to expand the *North Carolina.gdb*
2. Click the + sign to expand the *CensusData* feature dataset

QUESTION 7.12

Which of the following feature classes is NOT part of this feature dataset?

a. Landslides
b. Nctrct
c. USCounties
d. Ncblkgrp

Most of data we have examined up to this point focus on demographics (census geography and attributes) and at-risk populations, such as nursing home residents. There is a physical geography side to the fracking issue as well, however. Fracking is intimately tied to the earth's geology, so you will also need datasets that examine parameters related to this.

1. Create a new feature dataset in the *North Carolina.gdb* by right-clicking the geodatabase and choosing *New Feature Dataset*
2. In the dialog box that appears, type *NC_Geology* as the name and click *Next*
3. Click to select *NAD_1983_State_Plane_North_Carolina_FIPS_3200*
4. Click *Next*, then *Next* again to bypass the *Vertical Coordinate Systems* screen
5. Click *Finish*
6. Right-Click the *NC_Geology* feature dataset, and click *Import*, then *Feature class (multiple)*
7. Click the folder to the right of the *Input Features* box to navigate to your North Carolina geodatabase; double-click
8. Press and hold the *Ctrl* key on your keyboard while selecting the *Geology, Geology_Dikes*, and *Geology_Faults* shapefiles (*.shp)
9. Click *Add*
10. Verify that the *Output Geodatabase* is *NC_Geology*, and then click *OK*

You can watch the scrolling progress of the operation by choosing *Results* from the *Geoprocessing* menu and examining *Messages* under the listed operation.

11. From the *Help* menu, choose *ArcGIS Desktop Help*
12. Under the *Contents* panel, expand *Geodata*, then *Data Types* and *Feature datasets*
13. Read the overview

QUESTION 7.13

All feature classes in the same feature dataset must share a common coordinate system, and x, y coordinates of their features should fall within a common spatial extent.

a. True
b. False

14. In the *Catalog Tree*, right-click each of the original files left after you imported them into the new feature dataset and choose *Delete* to remove them

Raster Data

The raster data model represents the environment by dividing it into a number of discrete, coded cells that make up a grid format. Each cell in the grid is coded with a value that represents a specific quality or quantity of that location. These values can be codes, such as the primary land use occurring in the cell, or continuous numerical values like the elevation or temperature measured for each cell. The size of the individual cells, as well as their bit-depth (the number of bits used to encode the information in the computer), play roles in how much detail can be recorded.

In many instances, aerial photographs, in the form of raster data, are included in projects to use as background reference material when creating presentations of your analyses to the public.

1. Click *OakRidge_Subset.tif* to select it
2. Click the *Preview* tab

This file is an aerial photograph of Oak Ridge, NC.

3. Click the *Description* tab to answer the following questions:

QUESTION 7.14

Under *Resource Details*, what is the spatial representation type of file?

a. Polygon
b. Grid
c. TIN
d. Point

QUESTION 7.15

Under *Spatial Reference*, the ArcGIS coordinate system type is listed as projected.

a. True
b. False

4. In the *TOC*, right-click the file and choose *Properties*

QUESTION 7.16

What is the pixel depth of this file?

a. 12 bits
b. 4 bits
c. 32 bits
d. 8 bits

5. Select the *Preview* tab and zoom in to an area with a parking lot and cars
6. Continue zooming in until you can see the individual pixels

QUESTION 7.17

An individual car in the photograph is represented by less than a pixel.

a. True
b. False

7. Exit **ArcCatalog**

Chapter 8

Basic Spatial Analysis

The vector and raster data structures discussed in Chapter 7, along with any associated attribute data, form the basic inputs of GIS. When the data embedded in these structures are used to examine the geographic features they represent, then the GIS is being used to conduct spatial analysis. *Spatial analysis* is the process of studying geographic features on the basis of their locations. This can be as simple as selecting the features on the basis of some shared characteristic, to studying how multiple geographic features relate to each other spatially. Many of these techniques did not originate with GIS, but stemmed from the use of maps before the advent of computer technology. One of the earlier examples, for instance, occurred during the American Revolution. At the Battle of Yorktown in 1781, French cartographer Louis Alexandre Berthier produced a set of maps that depicted the positions of the British troops over time. He did this by hinging the maps together to produce *overlays* that could be flipped through to "see" temporal changes (Foresman, 1998).

Overlays are used to store geographic data as multiple layers of information. This technique can be used, as Berthier did, to visualize changes in the movement of features, or it can be used to combine, visually or analytically, two or more overlays that store different geographic data occurring in the same location. When used in the latter context, the process is known as *overlay analysis.* Widely touted as a classic example of pre-GIS overlay analysis is the map Dr. John Snow made in 1854. Snow's map is popular because it is an example of how a problem was solved using geographic relationships. In an effort to find a possible cause for a cholera outbreak in London, he mapped cholera deaths in relation to well locations; the graphic result allowed him to hypothesize that there was a link between the two. When he convinced officials to remove the pump handle from the well closest to the most deaths, the outbreak was contained, proving the link.

The *Canada Geographic Information System (CGIS)* marks one, if not the earliest attempt to automate the overlay process. Their use of computer technology marks the beginning of a revolution in geographic analysis. Today's GIS offers an abundance of analytical techniques, ranging from simple queries of attributes to more complicated combinations that include both attribute and spatial queries, as well as techniques that create new data.

Data Queries

Data queries are an essential aspect of GIS; they provide the mechanism for users to request information about spatial features. Different from *geoprocessing*, which generates new data from combining data already stored in the GIS, queries are used simply to retrieve already stored data about feature locations. They do not change the underlying data already present or create new data. Data queries can be approached from two distinct perspectives: *attribute queries* return information about feature attributes, and *spatial queries* return information about feature locations.

ATTRIBUTE QUERIES

Attribute queries select records of spatial features in a database using attribute values as the criteria for selection. An example might be to request that the GIS select and return just the counties of the United States in which the estimated housing value is greater than or equal to $75,000. Such queries are typically expressed using SQL (Structured Query Language), or some variant of SQL. SQL was designed specifically to manipulate data in relational databases. It uses attribute fields in the GIS database in conjunction with user-defined values for that field and set algebra ($<, >, =, <=, >=$) and/or Boolean algebra (AND, OR, NOT) to structure the request. In the example posed above, if the field in the database is [EST_HOMEVAL], then the expression needed to satisfy the request would be: [EST_HOMEVAL] >= 75000.

Those new to GIS, while most likely comfortable with set algebra, may not be as familiar with Boolean algebra. Boolean algebra is used to design more complex, or composite attribute queries. The most common Boolean operators in GIS include AND, OR, XOR, and NOT. AND is used when two conditions must both be true to select a feature, while OR selects features when either of the two conditions is true. XOR selects features when one condition is true, but the other is false, and NOT is used when a condition must be negated for the feature to be selected. Figure 8.1 provides a visual of how these functions work. If the request to the GIS was to select just the counties of the United States that belong to the state of Colorado in which the estimated home value is equal to or greater than $75,000, for example, then it becomes a composite query that requires the Boolean operator AND: [STATE_NAME] = 'Colorado' AND [EST_HOMEVAL] >= 75000 (Figure 8.2). If, instead, the request was to return just the counties of the United States that belong to the state of Colorado in which the estimated household value is over $75,000 or the level of educational attainment is more than four years of college, the query increases in complexity even more. It now requires not only the AND operator, but also the OR: [STATE_NAME] = 'Colorado' AND ([EST_HOMEVAL] >= 75000 OR [EDUCATION] >= 4). Notice the addition of the parentheses around the OR clause; queries are treated as mathematical statements, so they are evaluated from left to right using the mathematical principles of PEMDAS, which outlines the order in which one should evaluate each mathematical component (parentheses, exponents, multiplication, division, addition, subtraction). If the parentheses had been left out of this expression, then the GIS would have first selected all the counties in Colorado with the requisite estimated home value, then that subset or any county in the U.S. meeting the educational requirement would have been added (Figure 8.3).

The NOT operator is most likely the easiest to comprehend, the XOR the most difficult. If, for instance, the GIS was requested to return all the counties in the United States except those that have estimated home values less than $50,000, the expression required would be NOT ([EST_HOMEVAL] < 50000). If the expression loaded was [EST_HOMEVAL] < 50000 XOR [EDUCATION] > 4, then only those counties in which one condition is true while the other condition is false are selected.

FIGURE 8.1
Common Boolean functions.
Source: Elisabeth Nelson

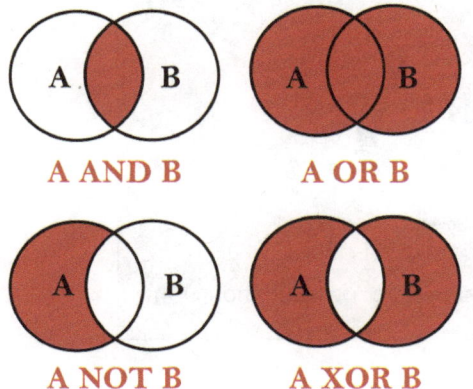

FIGURE 8.2
Composite query requiring the AND operator. Data: *Made with Natural Earth. Free vector and raster map data @ naturalearthdata. com; U.S. Census Bureau.*
Source: Elisabeth Nelson

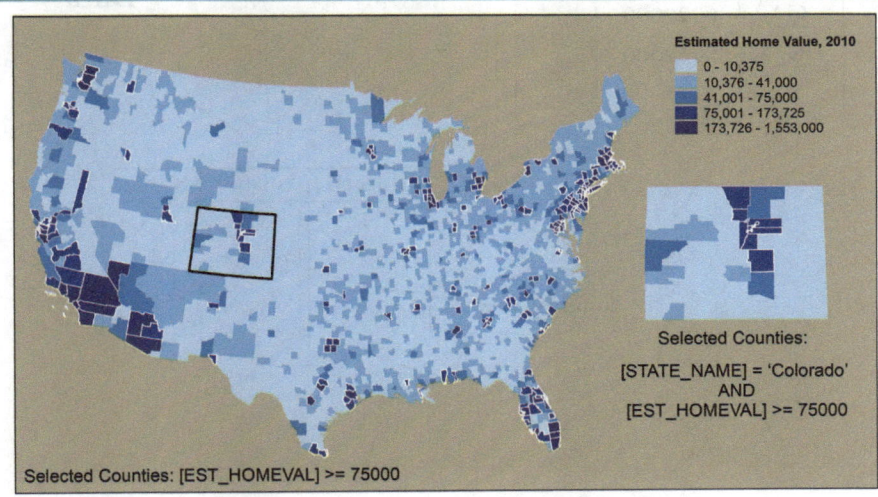

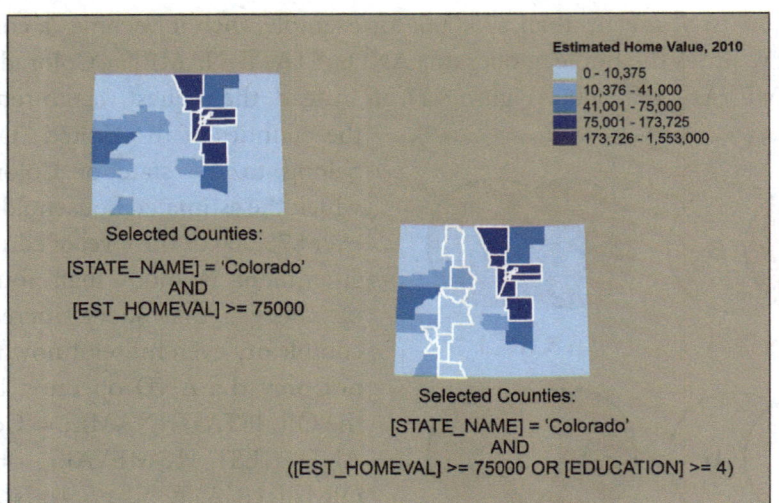

FIGURE 8.3
Complex query
requiring both the AND
and OR operator. Data:
*Made with Natural
Earth. Free vector and
raster map data @
naturalearthdata.com;*
U.S. Census Bureau.
Source: Elisabeth Nelson

In many cases, the feature attribute table will come with several attributes, but there is always the chance that the data you want to analyze or map is not among those. Since we want a flexible system, the GIS needs a way to incorporate 'outside' attribute tables. These tables are *non-spatial*; they are not tied to the spatial database. The challenge then, is to find a way to connect the table to the GIS databases. This is usually accomplished by 1) defining a common field between the non-spatial table and the feature attribute table, then 2) using that field to *join* the two tables. The result of a join is a virtual link to the non-spatial table via the feature attribute table. Once the tables are joined (Figure 8.4), you can then query and analyze that data as well.

FIGURE 8.4
Using a common field to join extra attribute data to a geographic database virtually.
Source: Elisabeth Nelson

Non-Spatial Data

Name	Population
GA	9920000
NC	9752000
SC	4724000
⋮	⋮

Spatial Data

ObjectID	Shape	Name
1	Polygon	GA
2	Polygon	NC
3	Polygon	SC
⋮	⋮	⋮

Join Key

SPATIAL QUERIES

It is also possible to query the database spatially. Spatial queries select features using the geometric characteristics of the database; features from one GIS layer are selected using features from a second layer. The spatial relationships between the two layers are the basis for selection. You might, for example, select all the parks in a parks layer that are within 30 miles of a school in a schools layer. This option is classified as *selecting features that are within a specified distance of another feature* and is an example of using the spatial relationship of *proximity* to answer a query. In ArcGIS, such queries are built using a dialog box that requires a *target* layer that is connected to a *source* layer by using a *spatial relationship* of interest. The target layer holds the spatial features to be selected; the source layer contains the features to which the target layer features are being compared. A spatial query using the *Intersect* option, for instance, would be looking for overlapping areas of interest between two layers; it might be specified as selecting *Counties that intersect Interstates* (Figure 8.5).

FIGURE 8.5
Spatial query in which counties are selected on the basis of their intersection with a selected interstate. Data: *Made with Natural Earth. Free vector and raster map data @ naturalearthdata.com.* Source: Elisabeth Nelson

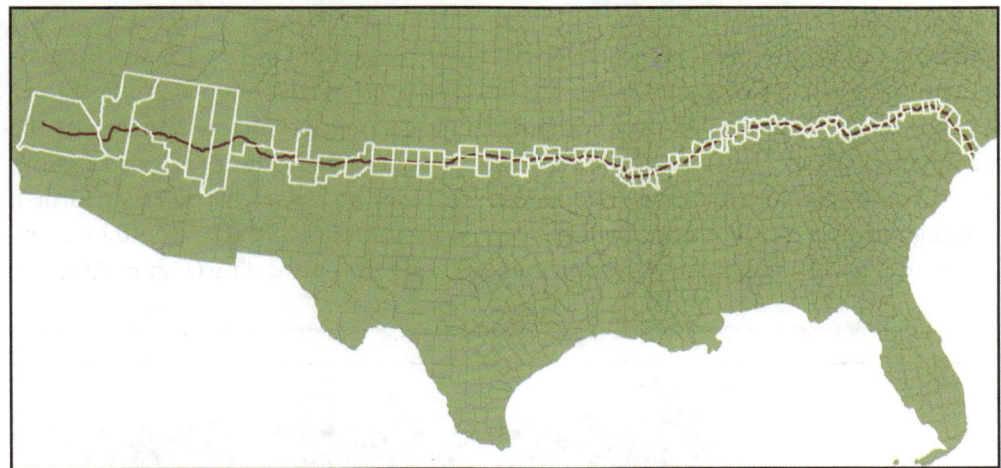

The Intersect relationship is an example of using the spatial property of intersection to answer a query. *Proximity* and *Intersection* are two of four basic spatial relationships commonly used to query features in a GIS; the other two are *Containment* and *Adjacency*. A containment query can be specified using multiple relationship descriptions; a common example might be to select the *Elementary schools that are completely within New Hanover County*. An example of using the adjacency relationship might be to select the *Counties that touch the boundary of New Hanover County*.

Geoprocessing Functions

Geoprocessing refers to those functions of a GIS that produce new data from the original data sets. Many methods fall under the geoprocessing umbrella, and they range from simple buffer operations to more complex statistical analyses. The most common of these include those that *extract* subsets of features from the original data, *aggregate* original features into larger features, *buffer* original features, and create new data by *overlaying* multiple data layers. More often than not, the question that prompted the analysis requires some subset of functions to be implemented sequentially to provide an answer.

EXTRACTING SPATIAL FEATURES

The *Clip* tool is used when the features in one data layer extend beyond the boundaries a project is set up to analyze. A data layer that serves as the outer boundary of the project is used to extract the required information from the layer covering the larger extent. From an anecdotal perspective, this can be thought of as using a cookie cutter to cut out a section of a data layer. The process overlays the outer boundary layer and the target feature layer, then computes their geometric difference to extract the data lying within the extent of the clipping boundary. The result of the process is a new feature class in which the area outside the outer boundary has been discarded, leaving only the data bounded by that polygon. A common example would be using the boundary of a county to clip point data covering the entire state (Figure 8.6).

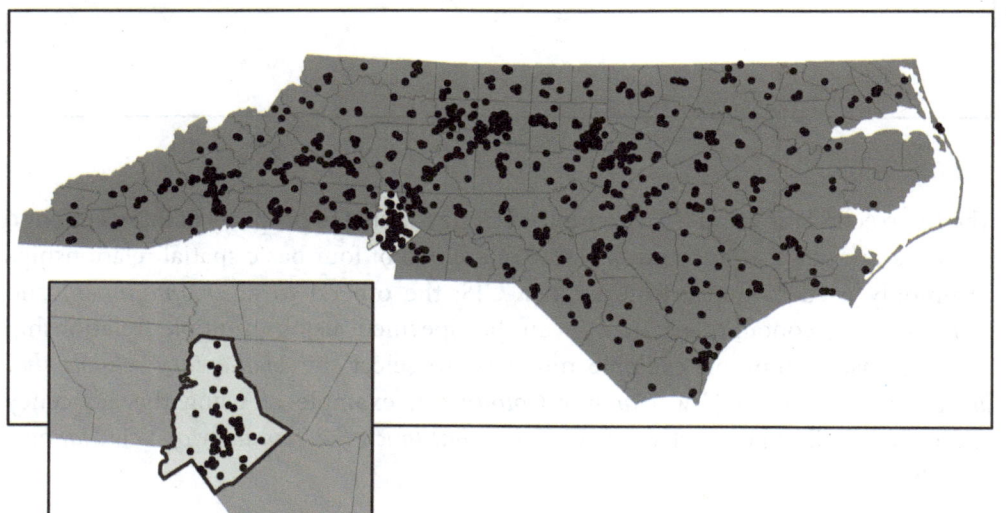

FIGURE 8.6
Clipping point features to restrict the selection to those within one county. Data: *Made with Natural Earth. Free vector and raster map data @ naturalearthdata.com; Department of Information Services; Guilford County, North Carolina, http://www.co.guilford. nc.us/departments/gis/.* Source: Elisabeth Nelson

AGGREGATING SPATIAL FEATURES

Dissolve. The *Dissolve* tool in a GIS creates new map features by collapsing adjacent area or line features in a data layer. A form of cartographic generalization, these features must share a common attribute value for the dissolve to be implemented. Features that are adjacent and have the same attribute values are generalized into a common larger polygon. If, for example, you needed a data layer showing the census regions of the United States, but only had a comprehensive U.S. states data layer, you could use an attribute that identifies each state's region to dissolve state boundaries on that basis. When complete, a new data layer organizing the states into larger census region polygons would be created. The capability to dissolve similar, adjacent polygons can greatly simplify a database by removing unnecessary boundaries and duplicate attributes.

Merge. Where the *Dissolve* tool collapses adjacent features in a data set, the *Merge* function works by combining multiple data layers into a single output layer. The input layers can be points, lines, or polygons, as long as all the data layers involved are the same data type. When the two data layers are adjacent, this is similar to tiling, which is the primary purpose of the tool. If, for example, two teams were assigned adjacent areas of a park and tasked with collecting GPS points outlining the trails of their section, a merge could be later used to combine the two files into one comprehensive layer of data.

BUFFERING SPATIAL FEATURES

The concept of buffering refers to creating a zone around specified spatial features. This zone, which can be either of constant or variable width, is then used to select a second set of features that surround the buffered features. Points, lines, and polygons can be buffered in a GIS. When point data are buffered, for example, a circle whose radius is the specified buffer distance will be created. These are saved to a new polygon file, where the polygons create an *inside region* and an *outside region*. The inside region, or area that is within the specified buffer distance represents the *buffer zone*, and the outside region represents the area that is more than the specified buffer distance from the feature.

Buffering applications often have the purpose of protecting some area, such as creating drug-free buffer zones that can be drawn around schools. Hospitals can also be buffered, if for example, there is a need to determine how many in a region are within one mile of a major transportation artery. A feature can also have more than one buffer zone associated with it; a point source of pollution, for instance, could be buffered with multiple distances as part of an assessment of how it influences wetlands (Figure 8.7).

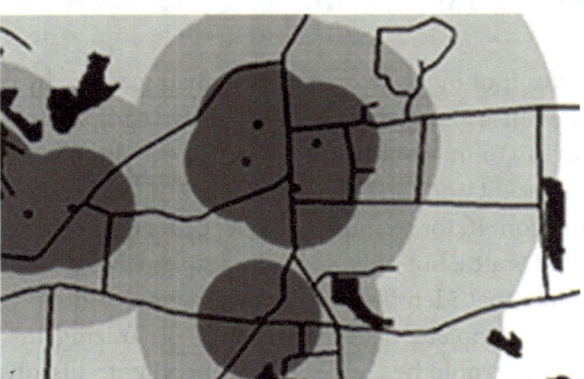

FIGURE 8.7
Multiple buffer rings
around point sources of
pollution. Image: *U.S.
Geological Survey, De-
partment of the Interior.*

OVERLAYING SPATIAL FEATURES

From the earliest implementation of GIS with the *Canada Geographic Information System*, users wanted to compare two maps of the same region analytically. An example of this might be the need to overlay a city's boundary file with a county's land use file. Prior to the development of GIS, cartographers created composite maps by overlaying two or more separate maps on a light table to visualize the result. To automate the process and create new spatial features resulting from the overlay, a combination of algorithms and Boolean algebra were used to join or fuse separate data layers as a function of the space they occupied. When working with vector data, these overlays allow polygon features to be combined with a second data layer of point, line, or polygon features. Different combinations of layers produce different results. For *Polygon-on-Polygon* overlays, the two most common ways of combining the input layers is by *intersecting* or *unioning* the data sets.

Intersect. The function of the *Intersect* command in a GIS is to select those areas that two data layers have in common spatially and create a new data layer on the basis of those features. This overlay uses the Boolean AND operator to determine the geometric intersections of all the polygons from both layers (refer to Figure 8.8). The polygons in the new data layer are then reconfigured geometrically by recalculating the coordinates, areas, and centroids of each to reflect the new polygons created, and each is tagged with all the attributes from the original layers. In our example above, intersecting the city boundary with the county land use would result in a new file containing the city boundary and just the land use polygons within that boundary. Where the extents of the two input layers differ, those polygons within one extent but not the other are eliminated.

Union. The *Union* command, on the other hand, takes the two input layers and combines all the polygons from both using the Boolean OR operation (refer to Figure 8.9). In this instance, the extents of both layers are combined. The polygons that are within either of these extents are kept in the new data layer. They are reconfigured geometrically in the new layer, with the attribute data combined as necessary. Looking at the city boundary and land use data from this perspective, all the polygons from both layers would be kept, with the spatial and attribute data recalculated to match the new configuration.

FIGURE 8.8

Intersection of county land use polygons with a city limits polygon. Data: *Department of Information Services; Guilford County, North Carolina, http://www.co.guilford.nc.us/departments/gis/.* Source: Elisabeth Nelson

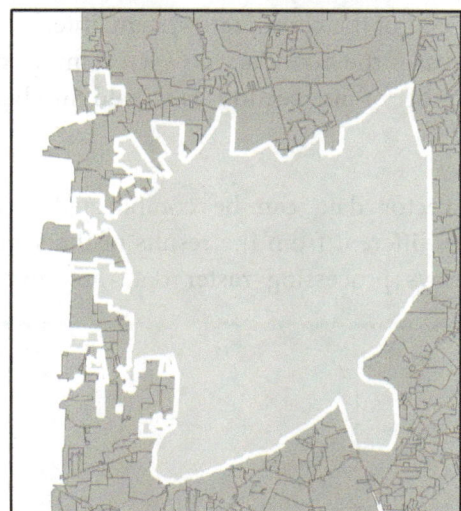 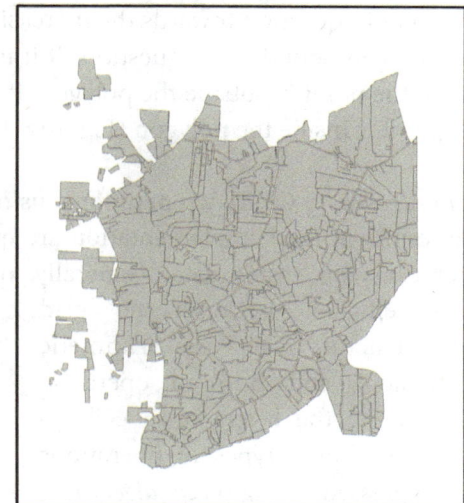

FIGURE 8.9

Union of county land use polygons with a city limits polygon. Data: *Department of Information Services; Guilford County, North Carolina, http://www.co.guilford.nc.us/departments/gis/.* Source: Elisabeth Nelson

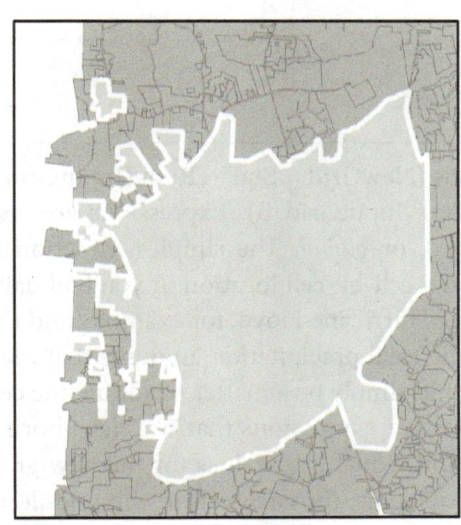 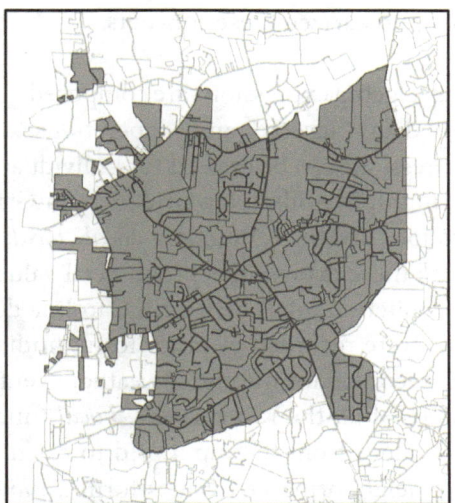

Point-in-Polygon. Points and polygons, as well as lines and polygons, can also be overlaid. The *Point-in-Polygon* method makes use of a *spatial join* to append the attributes from a polygon layer to a points layer, where each point is attributed with the attributes from the polygon surrounding it. When a target layer of point features, such as schools, are overlaid onto a join layer of polygon features, like census tracts, the result is a new point features layer in which each school has all the attributes of the surrounding census tract appended to its attribute table. This might be useful if you wanted to know something about the population living near the school, such as median family income, or median housing values.

The algorithm used to determine which polygon surrounds a point falls into the category of computational geometry. One common solution is to count how many times a line, parallel to one of the axes of the coordinate system and extending from the point in question towards the increasing end of the coordinate system, intersects the polygon boundary in question. If it is crosses the boundary an even number of times, the point is outside the polygon; if number of intersections is an odd number, the point is inside the polygon (Figure 8.10).

Map Algebra. Overlay operations using vector data can be complicated, and the results of their implementation are quite different from the results of the same operations on raster data. Generally, overlays processing raster data are much more efficient because the grid data structure can use arithmetic expressions and Boolean operators to combine the layers on a cell-by-cell basis. These types of operations are expressed using *map algebra*, a structure developed by Dana Tomlin (1990), who created a raster software package and accompanying textbook that popularized these concepts.

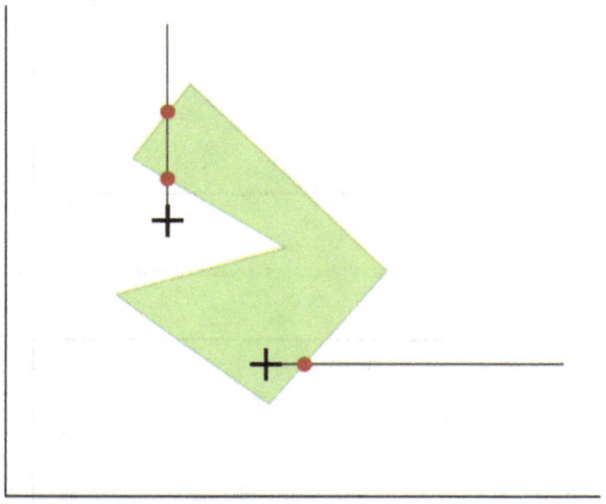

FIGURE 8.10
Solving the point-in-polygon request.
Source: Elisabeth Nelson

Map algebra expressions are composed of data layers and operations; expressions can be created for individual grids (NewGrid = SquareRoot (InputGrid)) or multiple grids (NewGrid = InputGrid A + InputGrid B). Expressions are also grouped by their function: *local*, *focal*, *zonal*, or *global*. The simplest function is *local*; these expressions process cell values on a cell-by-cell location. If you had daily precipitation totals for North Carolina during Hurricane Floyd, for example, and the data were recorded in raster form, finding the total precipitation amounts that each cell recorded for the entire weather event would simply be a matter of adding the cell values of all the raster layers. *Focal* functions are expressions that use neighboring cell values from the input grid to calculate individual grid values for the new grid. The neighboring cell values form a *moving window*; for each cell in the middle of that window, an operation can be performed that uses the surrounding cell values to modify the middle cell. If you have an elevation grid, for instance, you could use surrounding cell values to average each cell as you move across the surface, essentially smoothing out variation in elevation differences (Figure 8.11). Zonal functions use zones of cell values in calculating new grid values, and global functions take into account all cells in the grid in calculating each new grid value.

One useful way to explore the effects of raster overlays is to examine overlay results when using grid values of 0 (absence of feature) and 1 (presence of feature) only. As an example, let's work with two raster layers, one showing the presence of sparsely populated areas and the other showing the presence of geologically stable areas. Using these layers to help site a hazardous waste facility, the idea would be to find a location in a sparsely populated, geologically stable area. Thus, when the two layers are combined, the cells of interest should meet both criteria. If we multiply the two

FIGURE 8.11

Example of a focal function, in which a moving window is used to average cell values.
Source: Elisabeth Nelson

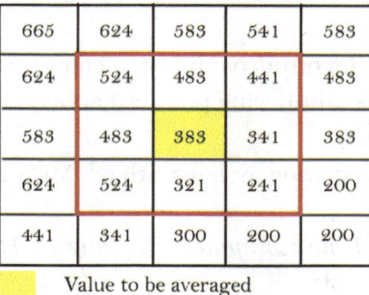

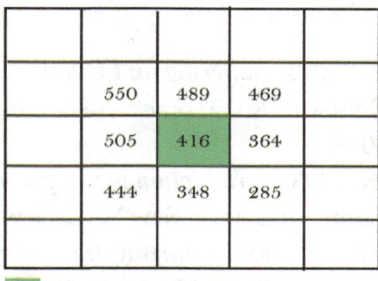

data layers, the data values for the new data layer would be either 0 or 1, where 1 would identify all the cells that met both criteria (Figure 8.12).

If, on the other hand, these constraints were relaxed so that only 1 of the 2 criteria had to be met, multiplying the two rasters wouldn't provide enough detail since the 0 category also includes cells that meet neither of the criteria. In this case, adding the two raster layers would be the more appropriate choice. When adding, the cells in the new raster layer would take 1 of 3 values: 0, where neither of the criteria are met, 2, where both of the criteria are met, and 1, where 1 of the 2 criteria were met. Of course, you could consider cells with a value of 2 as well as 1 as being valid in this scenario. Another way of approaching these results would be to consider the values to consist of graded choices, where 2 would identify the most desirable areas, 1 would identify less desirable areas, and 0 would identify those areas least desirable.

FIGURE 8.12

Example of a raster overlay in which the rasters are multiplied to mimic the AND Boolean function when binary values are used.
Source: Elisabeth Nelson

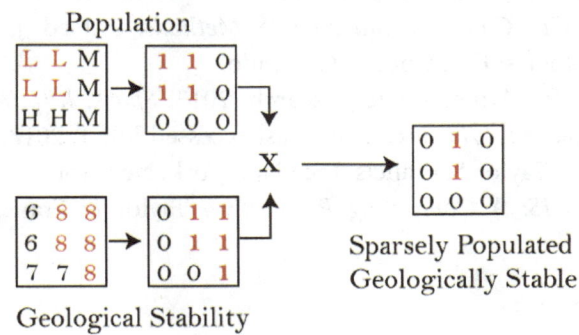

Additional Readings

Bolstad, P. 2013. *Buffering and Overlay in ArcGIS*. http://paulbolstad.cfans.umn.edu/Courses/FR3131/Lab%20Handouts/L9%20Buffering.pdf Last accessed July 2, 2013.

Chang, K. 2008. *Introduction to Geographic Information Systems*, 5th ed. McGraw Hill Companies, Inc.: New York, New York.

Chrisman, N. 2006. *Charting the Unknown: How Computer Mapping at Harvard Became GIS*. Esri Press: Redlands, California.

Clarke, K. 2003. *Getting Started with GIS*, 4th ed. Prentice Hall: Upper Saddle River, New Jersey.

Davis, B. 2001. *GIS: A Visual Approach*, 2nd ed. Onword Press: Albany, New York.

DeMers, M. N. 2000. *Geographic Information Systems*, 2nd ed. John Wiley & Sons, Inc.: New York, New York.

Environment Canada. 1973. The Canada Geographic Information System. *Cartographica*, v. 10(3): 62-86.

Escobar, F. 2013. *GIS Overlay Operations*.http://www.geogra.uah.es/patxi/gisweb/VOModule/VO_Operations.htm. Last accessed July 3, 2013.

Foresman, T.W. (ed.) 1998. *The History of Geographic Information Systems: Perspectives from the Pioneers*. Prentice Hall: Upper Saddle River, New Jersey.

National Academy of Sciences. 2006. *Beyond Mapping: Meeting National Needs Through Enhanced Geographic Information Science*. National Academies Press: Washington D.C.

Schmandt, M. 2013. *GIS Commons: An Introductory Textbook on Geographic Information Systems*.http://www.giscommons.org Last accessed July 2, 2013.

Slocum, T., McMaster, R., Kessler, F., and H. Howard. 2008. *Thematic Cartography and Geovisualization*, 3rd ed. Prentice Hall: Upper Saddle River, New Jersey.

Theobald, D.M. 2007. *GIS Concepts and ArcGIS Methods*, 3rd ed. Conservation Planning Technologies: Fort Collins, Colorado.

University Corporation for Atmospheric Research. 2013. *Spark. http://eo.ucar.edu/staff/dward/Lisa%203100/lecture5.doc3*. Last accessed July 1, 2013.

Wise, S. 2002. *GIS Basics*. Taylor & Francis, Inc.: New York, New York.

Worboys, M. F. 1995. *GIS: A Computing Perspective*. Taylor & Francis: London, England.

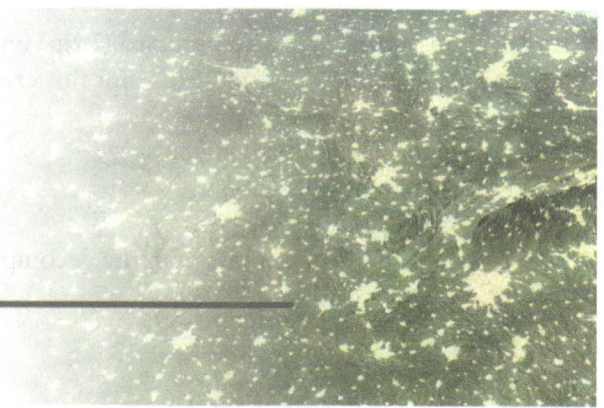

Exercise 8

Solving Spatial Problems

In exercise 8, you will analyze your data from Chapter 7 to assess the potential impacts of fracking in North Carolina, Guilford County, and the Town of Oak Ridge.

OBJECTIVES

- Subset a spatial data layer to create a new layer of more limited extent
- Dissolve features to create a more generalized layer of data
- Buffer a spatial data layer
- Overlay two polygon layers
- Perform a point-in-polygon analysis

SOFTWARE INFORMATION

Introducing ArcMap uses Esri's **ArcGIS 10.6** software. **ArcGIS** is a commercial GIS package, available in geospatial labs affiliated with schools that have a campus-wide site license for the software. Instructors at these campuses may also request 1-year student versions of the software at http://www.esri.com/landing-pages/education-promo.

DATA

The data for the exercises is available from the Kendall Hunt Student Ancillary site. See the inside front cover for access information. You may also be directed to download the data from a different location by your instructor.

1. Download the data as instructed by your instructor
2. Save the file to a location where you have read/write privileges (USB key, home computer, class server)

The file you just saved is in a compressed (*.zip) format, and was created using **7-zip** freeware (http://www.7-zip.org). To use the data for this exercise, you must decompress it using the same software.

3. Locate the zipped file that you saved
4. Right-click the file
5. Choose *7-Zip - Extract Files*
6. Click *OK* to create a folder with the decompressed data

Specify a Home Folder

In this exercise, rather than opening an already existing ArcMap map document, you will be creating one from scratch. Because of this, you will also need to explicitly associate a Home Folder and a default geodatabase for your new map. This requires a few steps up front, but will make things easier later.

1. Start **ArcMap** by double-clicking on your desktop icon or by clicking >All Programs > ArcGIS > **ArcMap 10.6**
2. In the *ArcMap - Getting Started* dialog box, click *Cancel*
3. If the *Catalog* window is not open, click on the *Catalog* tab on the right side of the ArcMap interface OR go to *Windows – Catalog* to open it
4. Click the pushpin in the upper right corner of the *Catalog* window to freeze the window in place (Figure 8.1)

FIGURE 8.1
Used with permission. Copyright © 2018 Esri, ArcGIS, ArcMap, Arc-Catalog, United States Department of the Census; naturalearthdata.com All rights reserved.

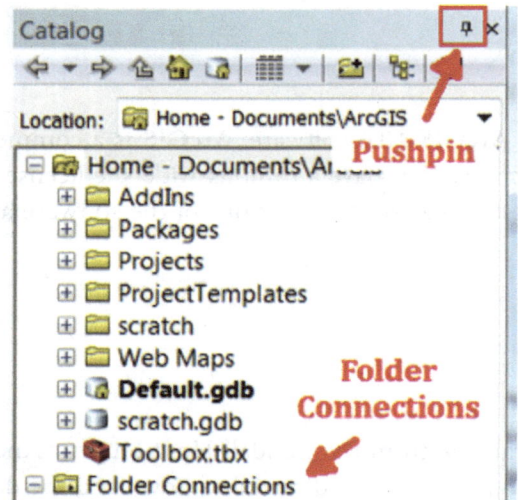

Notice that the default Home folder is set *Home-Documents\ArcGIS*. You want your Home folder to be *Chapter 08*, which is where your data for this exercise resides.

5. From File, choose *Save As*
6. Name your file using this convention: *lastname_Exercise08*
7. Save your map to the *Chapter08* folder containing *Exercise_08.gdb*

Your home folder should now be updated to the correct location. If you have any questions, verify this with your instructor before proceeding.

Specify a Default Geodatabase

It is also important to link your map document specifically to the geodatabase that houses your data for the exercise. With this link in place, any changes or additions you make to your data will be stored in the correct location on the computer.

1. From the *File* menu, choose *Map Document Properties*
2. Click to select the file folder button to the right of the default geodatabase path
3. Browse to find *Exercise_08.gdb*
4. Click to select it, then click *Add*
5. Next to pathnames, place a checkmark by *Store relative pathnames to data sources*
6. Click *OK*

Analysis Overview

The Town of Oak Ridge has two broad areas in which they have specific questions: water contamination potential and earthquake risk potential. These questions will be used to frame your analysis.

For water contamination potential, the questions they have are:
- Are any potential fracking sites within 5/10/20 miles of Oak Ridge?
- Which water supply watersheds overlap Triassic beds in NC?

For earthquake risk potential, their questions are:
- How many faults and dikes are in North Carolina, Guilford County, and Oak Ridge?
- How many at-risk sites are within a mile of these faults and dikes?

WHERE ARE THE POTENTIAL FRACKING SITES IN NC?

The *NC_Geology* feature dataset contains a layer, *Geology* which lists information on all the rock types in the state. One of the columns in the attribute table (BELT) will allow you to create a new layer of the geologic features that are likely to be drilled – these are the Triassic Basins.

1. In the *Catalog*, find your *Home* folder
2. Click the + sign to the left of *Exercise_08.gdb* to see the spatial data files you will use in this exercise
3. Click and drag the *Geology* feature class from the *Catalog* window to your *Data View* window
4. Right-click the *Geology* data layer in the *TOC* and choose *Open Attribute Table*

Each of the polygons in the table belongs to a particular geological belt: the belt in which fracking is likely to occur in NC is called *Triassic Basins*

5. Right-click the heading named BELT, and then click *Sort Ascending*
6. Scroll to the very bottom to verify that *Triassic Basins* are listed as a BELT value
7. At the top of the table, to the left, are a set of icons–scroll across these until you find *Select by Attributes* and click it (Figure 8.2)

FIGURE 8.2

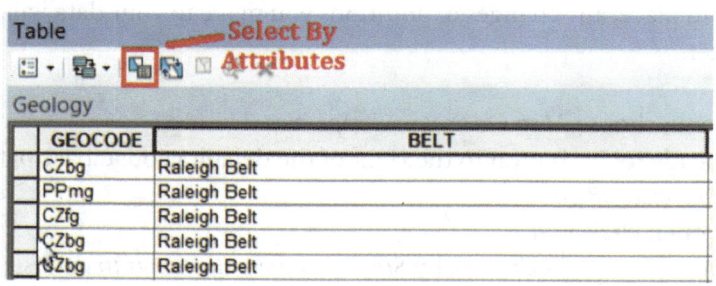

The Select by Attributes dialog allows you to query the database to elect a subset of records (Figure 8.3)

FIGURE 8.3

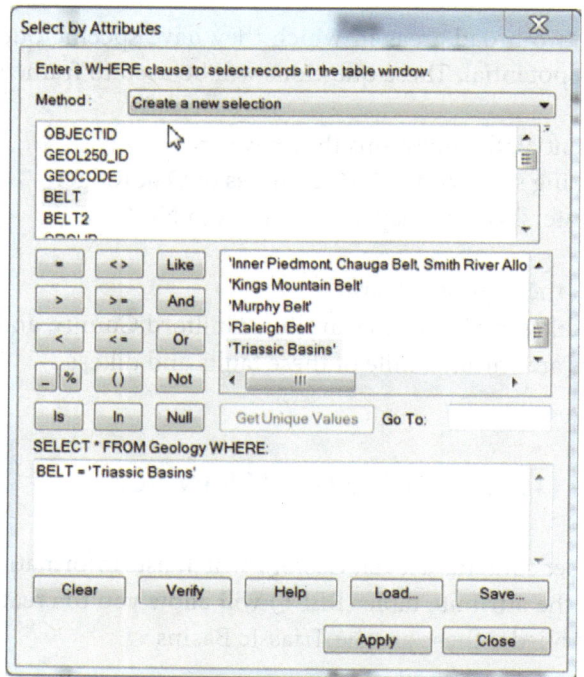

8. For *Method*, choose *Create a new selection* in the top drop-down box
9. In the upper window listing fields of the table, double-click *BELT*
10. Click the equals (=) button below that window
11. Click the *Get Unique Values* button
12. Scroll down to the bottom and double-click *Triassic Basins*

The query should read BELT = 'Triassic Basins'

13. Click the *Verify* button to validate the query expression, then *OK* to close the window if the query is successful
14. Click *Apply* then *Close*
15. At the bottom of the Table display, locate and click on the *Show Selected Records* icon (Figure 8.4)

FIGURE 8.4

Used with permission. Copyright © 2018 Esri, ArcGIS, ArcMap, Arc-Catalog, United States Department of the Census; naturalearthdata.com All rights reserved.

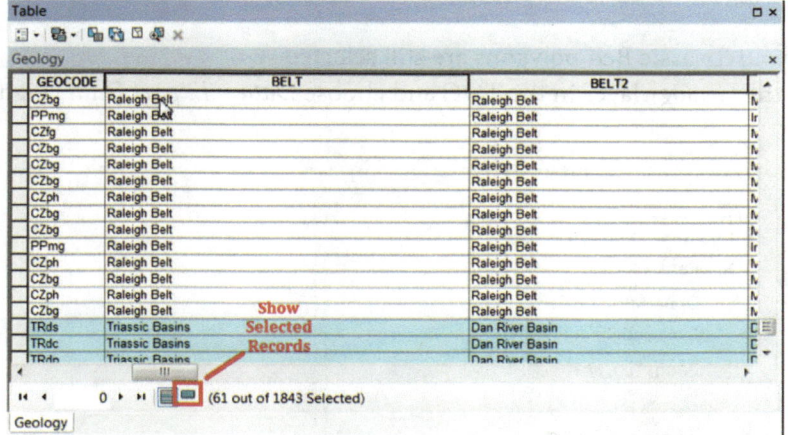

QUESTION 8.1

How many records were selected?

  a. 40
  b. 52
  c. 61
  d. 110

16. Close the table window
17. Zoom in to explore, then select *Triassic Beds* on your map

SAVING YOUR MAP DOCUMENT

1. From the *File* menu, choose *Map Document Properties*
2. Click the file folder button to the right of the default geodatabase path and browse to the *Exercise_08.gdb*
3. Click to select it, then click *Add*
4. Next to pathnames, place a checkmark by *Store relative pathnames to data sources*
5. Click *OK*
6. From *File*, choose *Save As*
7. Name your file using this convention: *lastname_Exercise08*
8. Save your map to the *Chapter08* folder containing *Exercise_08.gdb*

DISSOLVING THE POLYGONS

1. Make sure your Triassic Bed polygons are still selected
2. Right-click the *Geology* layer in the TOC and choose *Data – Export Data* (Figure 8.5)

FIGURE 8.5
Used with permission.
Copyright © 2018 Esri,
ArcGIS, ArcMap, Arc-
Catalog, United States
Department of the
Census;
naturalearthdata.com
All rights reserved.

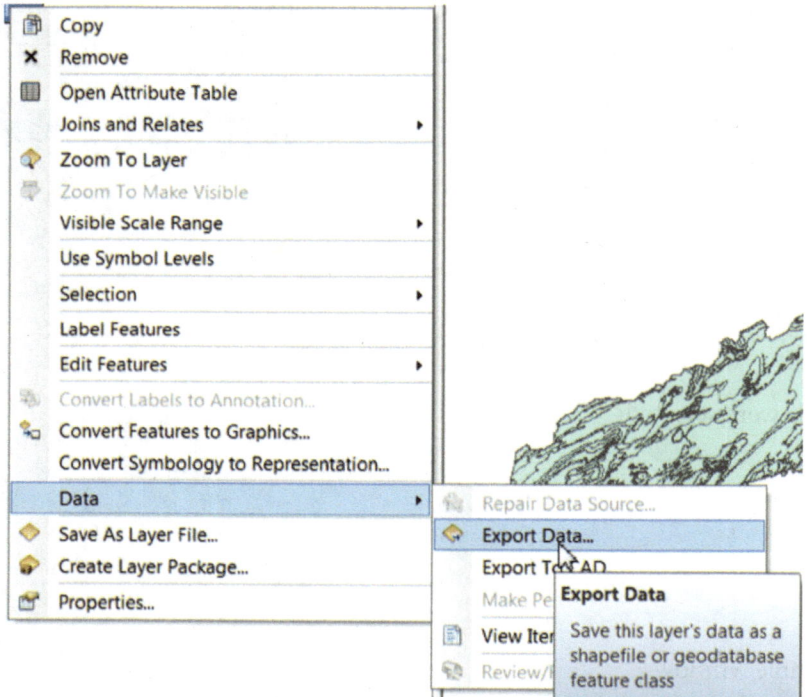

3. In the pull-down menu next to Export: choose *Selected features*
4. In the *Output Feature Class* box, navigate to your *Chapter08* folder
5. In the *Saving Data* dialog box, find *Save as Type* and choose *File* and *Personal Database Feature Classes*
6. Double-click *Exercise08.gdb*
7. Double-click the *NC_Geology* feature dataset
8. Save your new file as *Triassic_Basins* inside the dataset
9. Click *Save*, then *OK*
10. When prompted to add the new data layer into your map document, click *Yes*
11. From the Main Menu, click *Geoprocessing*, then *Dissolve* (Figure 8.6)

FIGURE 8.6

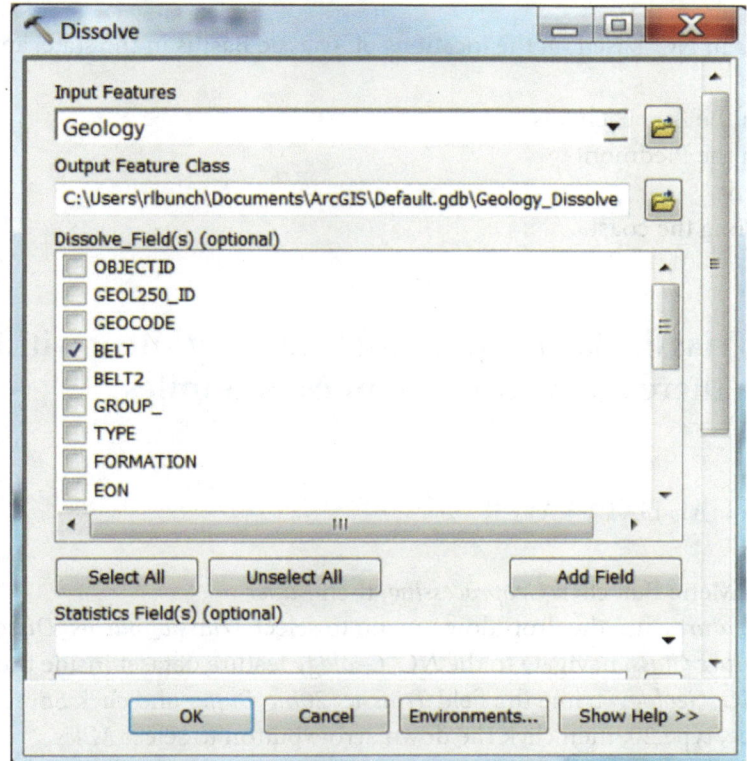

12. To the right of the *Input Features* box, click the down arrow
13. Select *Triassic_Basins*
14. For *Output Feature Class*, navigate to your *NC_Geology* feature dataset in *Exercise08.gdb*
15. Double-click *NC_Geology*, then name your file *Triassic_Basins_Dissolve* and click *Save*
16. In the *Dissolve_Fields* window, Check BELT, and then click *OK*
17. To check on the progress of your new output, go to the *Geoprocessing* menu and choose *Results*, then expand *Current Session-Dissolve-Messages*
18. Under the *Selection* menu, choose *Clear Selected Features*
19. f needed, change the symbology for the dissolved polygons so they stand out relative to the rest of the NC geology
20. Save your map document

QUESTION 8.2

How many features are listed in the attribute table of *Triassic_Basins_Dissolve*?

   a.   25
   b.   10
   c.   4
   d.   1

QUESTION 8.3

Potential fracking sites in NC, based on the locations of Triassic Basins in the state, appear:

   a.   to be located in the Appalachians
   b.   to be located in the Piedmont
   c.   to be nonexistent
   d.   to be located along the coast

Are any of the Triassic Basins (potential fracking sites) within 20 miles of Oak Ridge? Are there any within 10 miles or 5 miles?

BUFFERING A SPATIAL DATA LAYER

   1.   From the Main Menu Bar, click *Geoprocessing*, then *Buffer*
   2.   For the *Input Features*, use the drop-down menu to select *Triassic_Basins_Dissolve*
   3.   For *Output Feature Class*, navigate to the *NC_Geology* feature dataset inside the *Exercise08.gdb*
   4.   Double-click *NC_Geology*, name the field *Triassic_20mi_Buffer* and click *Save*
   5.   For the Distance, type 20, then click the down arrow button to select *Miles*
   6.   For *Side Type* choose *Outside Only*
   7.   If the dialog box's *Help* is not shown to the right, click the *Show Help button*

QUESTION 8.4

Side Type refers to the sides of the input features that will be buffered. The option OUTSIDE_ONLY will:

   a.   generate buffers around the polygon and will contain and overlap the area of the input feature
   b.   generate buffers along only the left topological side of the polygon
   c.   generate buffers along only the right topological side of the polygon
   d.   generate buffers only outside the area of the input feature

8. *For Dissolve type* choose *All*
9. Click *OK*
10. When the layer appears, right-click the layer in the TOC and choose *Properties*
11. Click the *Display* tab, and change the *Transparency* to 30%
12. Click the *Symbology* tab and change the buffer color
13. Click *OK*
14. In the *Catalog* window, click the + sign next to the *Guilford_Co* feature dataset in *Exercise_08.gdb*
15. Click and drag *Oak_ridge_Limits* to your *Data View* window

QUESTION 8.5

There are Triassic Basins (potential fracking sites) within 20 miles of Oak Ridge.

a. True
b. False

16. Repeat the buffering operation to create a 10-mile buffer around *Triassic_Basins_Dissolve*
17. In the TOC, click and drag the 10-mile buffer layer so that it is underneath the *Oak_ridge_Limits* layer
18. Right-click the layer and choose *Properties*
19. Click the *Display* tab and change the *Transparency* to 30%
20. Click *OK*, then Zoom in closer to the Oak Ridge city limits
21. Save your map document

QUESTION 8.6

There are Triassic Basins (potential fracking sites) within 10 miles of Oak Ridge.

a. True
b. False

22. Repeat the buffering operation to create a 5-mile buffer around *Triassic_Basins_Dissolve*

QUESTION 8.7

There are Triassic Basins (potential fracking sites) within 5 miles of Oak Ridge.

a. True
b. False

Which water supply watersheds overlap Triassic beds?

This section of the analysis examines the potential effects of fracking on water supplies.

CREATING A POLYGON OVERLAY

1. In the *Catalog* window, click the + sign next to the *NC_Water* feature dataset to view the feature classes related to water in NC
2. Click and drag the *Water_Supply_Watersheds* feature class to your *Data View* window
3. When the layer appears, right-click it in the *TOC* and choose *Properties*
4. Click the *Display* tab, and change the *Transparency* to 30%
5. In the *TOC*, make all the buffer layers invisible and click *OK*
6. From the Main Menu, click *Geoprocessing*, then *Intersect*
7. For *Input Features,* select *Water_Supply_watersheds* and *Triassic_Basins_Dissolve*
8. For the *Output Feature Class*, navigate to *Exercise08.gdb*, name the new layer *Triassic_Water_intersect* and click *Save* (Figure 8.7)

FIGURE 8.7

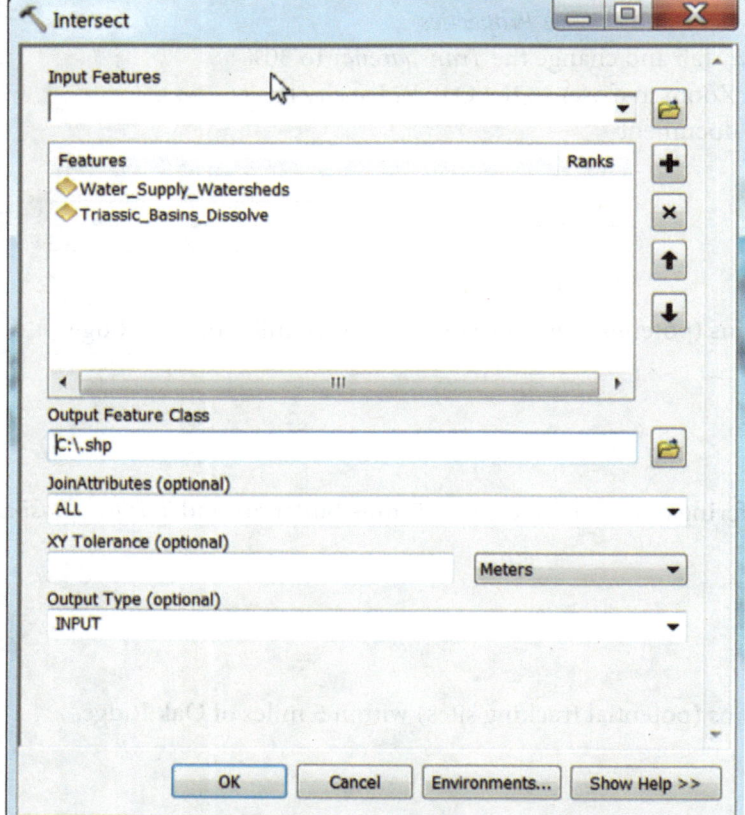

9. Click *OK* and zoom out to see the results
10. In the *TOC*, make *Water_Supply_Watersheds* invisible
11. In the *Catalog* window, find the *NC_Counties* feature class, located in the *NC_Geology* feature dataset, and click and drag it to your *Data View* window
12. In the *TOC*, click and drag the data layer to the bottom of the list of layers
13. Uncheck the *Geology* data layer to make it invisible
14. Right-click *NC_Counties* and choose *Label Features*

QUESTION 8.8

Use the *Identify* tool to explore the intersected polygons in Rockingham and Stokes counties. What is the name of the river basin that could potentially be polluted in these counties?

a. Yadkin
b. Neuse
c. Roanoke
d. Cape Fear

Are there faults and dikes in the study region?

The second series of questions the town of Oak Ridge wants to explore involves earthquake potential. The east coast of the United States is thought to be much more geologically stable than the west coast, but increasing the level of fluids in the underlying rock strata could act as a lubricant and allow areas that have previously been considered more stable to see an increase in earthquake activity. The first issue they would like you to address is whether any faults or dikes are in the region.

1. In the *TOC*, double-click the current data frame and under the *General* tab rename it: *Water_Impacts*
2. From the *Insert* Menu, insert a second data frame and name it *Earthquake_Impacts*
3. In the *Catalog* window, find the *NC_Geology* feature dataset in *Exercise08.gdb* and click the + sign next to it to expand it
4. Click and drag the following feature classes to your *Data View*: *Geology_Dikes, Geology_Faults, Triassic_Basins_Dissolve,* and *NC_Counties*
5. Add *Oak_Ridge_limits* from the *Guilford_Co* feature dataset in *Exercise08.gdb* to your *Data View*
6. Right-click *NC_Counties* in the *TOC* and choose *Label Features*
7. Zoom in to Guilford County – make sure this layer is at the bottom of the *TOC* so it does not cover up the other data layers
8. Change the symbology properties as needed to distinguish all your features (right-click each layer in the *TOC* and choose *Properties-Symbology*)

QUESTION 8.9

There are geologic faults that run through Guilford County in NC.

    a. True
    b. False

QUESTION 8.10

There are geologic dikes that run through Guilford County in NC.

    a. True
    b. False

QUESTION 8.11

There are geologic dikes that run through the town of Oak Ridge, NC.

    a. True
    b. False

CONDUCTING A POINT-IN-POLYGON ANALYSIS

1. From the *Catalog* window, add *NC_Nursing_Homes* from *Exercise_08.gdb* to your *Data View*
2. From the *Geoprocessing* menu, choose *Buffer*
3. Create a 1-mile buffer around the dikes in *Geology_Dikes*
4. For the *Dissolve_Type* choose *All*
5. Name the new layer *Dike_Buffer_1mi*
6. Click *OK*
7. From the main menu, click *Selection*, then *Select by Location* (Figure 8.8)
8. For *Selection Method*: Choose *Select features from*
9. For *Target Layer(s)*: check *NC_Nursing_Homes*
10. For *Source Layer*: choose *Dike_Buffer_1mi*
11. For *Spatial selection method for target layer features*: choose *intersect the source layer feature*
12. Click *OK*
13. Open the *Attribute Table* for the *NC_Nursing_Homes* data layer to see how many features are selected

FIGURE 8.8

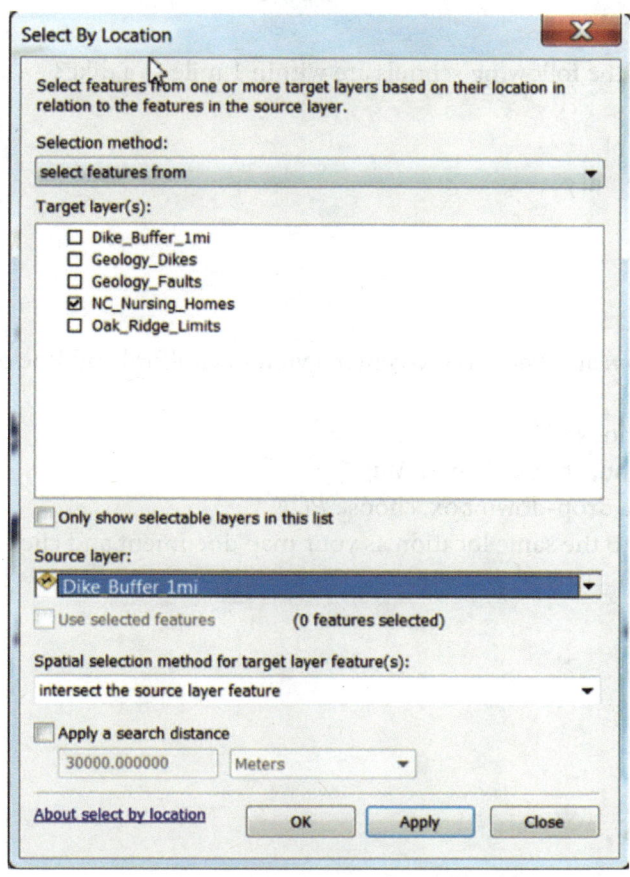

QUESTION 8.12

How many nursing homes in NC are within 1 mile of a dike?

a. 1,027
b. 7
c. 72
d. 59

14. Under the *Selection* menu, choose *Clear Selected Features*
15. Add *GC_Schools* (inside the *Guilford_Co* feature dataset) in the *Exercise08.gdb* to your data view
16. Repeat the point-in-polygon analysis for the *GC_Schools* data layer

QUESTION 8.13

In Guilford County, which of the following schools are within 1 mile of a dike?

    a.   Kernodle Middle School
    b.   General Greene Elementary
    c.   Page High School
    d.   Southwest Elementary

17. Use the *Data View Zoom* and *Pan* tools to center a view of Guilford and Rockingham counties in your *Data View*
18. From the *File* menu, choose *Save*
19. Now, from the *File* menu, choose *Export Map*
20. Under the *Save as Type* drop-down box, choose *PDF*
21. Save this map version to the same location as your map document and click *Save*
22. Exit ArcMap

Chapter 9

Aerial Photography

The foundation of modern mapping, as we saw in *Chapter 5: Mapping Location*, occurred with the development of those surveying techniques used in producing the survey of France, which the Cassini family completed in the late 1700s. As other countries began to implement France's new techniques for their own national surveys, the demands of this new standard in mapping quickly became evident. The sheer number of bodies, for example, required to carry out the surveys was a problem solved by using armies to complete the work. Involving the military, however, changed the perception of the need for these surveys. Before national surveys were associated with military applications, many countries, especially those lacking the financial and technological means to conduct them, didn't see a need for such detailed mapping within their borders (Dickinson, 1979).

This perception didn't truly begin to change until the advent of World War I (1914-1918). The reality of a war global in scale made accurate mappings of foreign lands critical. This is where the need for better equipment and improved technology also became issues. One problem in mapping foreign lands was accessibility; being able to produce or obtain a map of such an area tied to surveying technology was exceedingly rare. Most countries, at the time, resorted to *compilation*, or piecing a map of an area together by using one or more, usually inferior, sources (Dickinson, 1979). World War I, however, spurred developments in technology on several fronts, including expanding on the idea of combining photography with manned flight. Both the development of photography and the ability for man to fly occurred before the 20th century, but the full impact of *aerial photography* on mapping wasn't realized until well into that era.

Photography, for example, has as its foundation studies of the nature of light, which can be traced as far back as ancient Greece. It is the principles derived from these studies, along with advances in chemistry and optics, that all converged in the early 1800s to help *Joseph Nicéphore Niépce* produce the first known photograph. Taken in 1826, Niépce's photo is believed to show the courtyard outside of his house in France (Thrower, 1996).

From this inauspicious start, it was merely a matter of time before someone would conceive of photographing his or her location from above the ground. *Gaspard-Félix Tournachon* (also known as *Nadar*) was that man; in 1858, he used a camera from an anchored balloon to take a picture of the terrain of Biévre, France. This image, captured on glass, no longer exists. It is, however, the first record we have of an attempt to create an *aerial photograph*, or picture of a portion of our planet taken from a position above the surface of the earth. Nadar's work captured the imagination of the public, and the demand for aerial photography was officially born (Aronoff, 2005). Two short years later, Boston photographer *James Wallace Black* teamed up with balloon navigator *Samuel Archer King* to take the first aerial photograph in the United States. King took Black up in his hot-air balloon 1,200 feet over the city of Boston and Black took an image. This image, titled *Boston, As the Eagle and the Wild Goose See It,* is the oldest known aerial photograph that still exists (Lillesand, et al., 2007).

In 1862, the Union Army adopted this form of aerial photography to assess Confederate defenses during the Civil War (1861-1865). While useful in theory, the Army soon found it less than practical, as the balloons made easy shooting targets. What they needed was a sturdy platform that could take photos as it moved. The answer to this, unfortunately, was some decades away. The U.S. Army Balloon Corp disbanded in 1863, and it wasn't until *Orville* and *Wilbur Wright* unveiled their airplane in 1903 that technology provided a more suitable platform. Their first attempt at flight, on December 14, only lasted 3 seconds, but before the week was out, they were covering hundreds of feet at altitudes of up to 10 feet (Gray, 2002). By 1908, the first public flights had been witnessed both in the United States and Europe, and in 1909, while in Italy conducting flight training for the military, the first aerial photograph was taken from one of their planes (Aronoff, 2005).

During World War I, planes became a key component of military strategy and the continued development of air photo technology and interpretation became crucial. The first cameras developed specifically for aerial photography were in place by 1915, and skilled photo interpreters, the "eyes of the armed forces" (Aronoff, 2005:13), became a new career in high demand by the war's end. The mainstreaming of aerial photography and its associated photointerpretation methods revolutionized the civilian fields of surveying and map production. Those working in forestry, geology, agriculture, and cartography, among other fields, began to adopt *photogrammetry* as a way to reduce the costs and time required to produce and update large-scale maps.

Photogrammetric technology concerns itself with obtaining accurate measurements from aerial photographs and using those measurements to produce even higher-quality maps than traditional surveying, of both easy-to-access as well as difficult-to-access areas.

The Aerial Photograph: Key Characteristics

Using aerial photographs as the basis for producing large-scale maps, such as topographic quadrangles, fundamentally changed the mapping sciences. Understanding some of the basic characteristics of these images is a good place to start when exploring how they have influenced mapping. For example, a raw aerial photograph may never be treated as equivalent to a map. It is not selective or generalized, as a map is; even more importantly, though, is that there is distortion embedded in its geometrical foundation that is uncontrolled in the photograph's raw format. The field of photogrammetry, in fact, evolved from the need for surveyors to correct these distortions and extract real-world dimensions. A classic way to envision these types of problems is to examine the shape of a feature on an unprocessed aerial photo and compare it to its real-world shape on both an orthophotograph and a topographic map. In Figure 9.1, for example, the land cleared for power lines in the unprocessed photograph (the straight lighter-colored line running through the middle of the photo) bends and dips in places due to changes in elevation as it crosses the landscape (a). The changes in elevation recorded by the camera cause the scale of the photograph to vary, making it unsuitable for use as a map or as the basis for creating a map. To use the photograph as the foundation for creating a map, these types of distortions must first be removed, which results in an *orthophotograph* (b). The orthophotograph is the basis for creating the topographic map (c).

FIGURE 9.1
Unprocessed aerial photograph, orthophotograph and topographic map of the Altavista, Virginia area. The power line clearing appears to bend in the unprocessed photograph due to distortion caused by relief displacement. Photos and topographic map: *U.S. Geological Survey, Department of the Interior.*

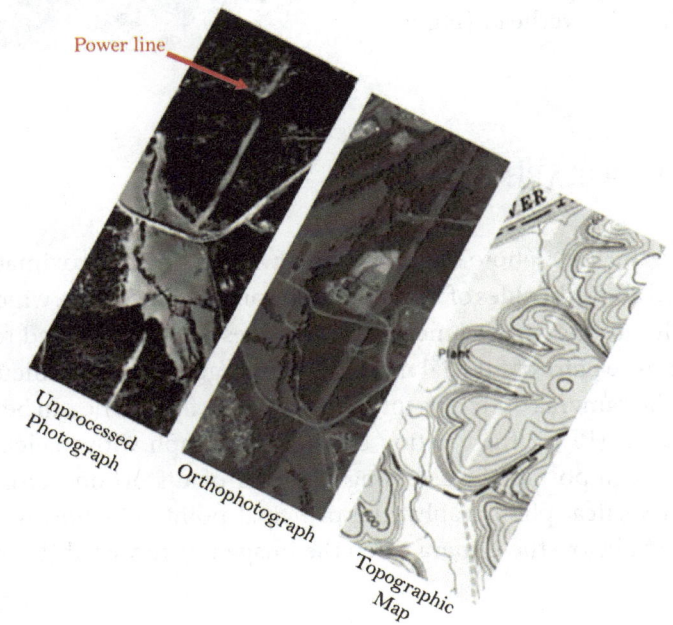

Power line

Unprocessed Photograph

Orthophotograph

Topographic Map

The following are some key characteristics of aerial photographs that influence these types of distortions.

TYPES OF AERIAL PHOTOGRAPHS

Aerial photographs can be grouped into two broad categories: *oblique* and *vertical*. Oblique photographs are probably the type most familiar to the general population. Taken when the axis of the camera is tilted at an angle to the ground during imaging, the result is a *perspective view* of an area, which mimics what we might see ourselves if we were viewing the same area at a similar height (Figure 9.2a). Unfortunately, the tilt of the camera further exaggerates the geometric distortion inherent in the image. Vertical photographs, on the other hand, are far less complicated to correct geometrically, a requirement for using them as the foundation for a map. These photographs are imaged with the camera pointed vertically to the ground. The result in this case is a *plan view*, in which the area is seen as if the viewer is directly overhead (Figure 9.2b).

a.

b.

FIGURE 9.2
Aerial photograph types: (a) Oblique view of Starmount area in Greensboro, NC, 1985; (b) Vertical view of Cardinal neighborhood in Guilford County, NC, 1989. Photos: *Department of Geography, UNCG.*

AERIAL PHOTOGRAPH FORMAT

The standard vertical air photo is a black and white print, approximately 9 inches square. Centered along the sides of the photograph are *fiducial marks*, which are images of markers built into the aerial camera. These images are used as fixed references; by connecting opposite pairs of fiducial marks on an image with perpendicular lines, the point at which the camera's optic axis intersects with the film plane can be determined. This *principal point* (PP) is the location on the photograph with the least geometric distortion, and is important in assessing patterns of distortion across the image (Figure 9.3). In vertical photography, the principal point coincides with the nadir, the point on Earth below the camera when the image was recorded (Monmonier and Schnell, 1984).

FIGURE 9.3

Fiducial marks and the
principal point on an
aerial photograph.
Photo: *Department of
Geography, UNCG.*

Most aerial photographs are imaged during a series of parallel, overlapping flight lines
the aircraft follows when taking the images (Figure 9.4). The resulting photographs
are configured into long strips, with the final prints oriented 90 degrees to the flight
line. Adjacent strips differ by 180 degrees in orientation, as the plane turns at the end
of each run. As flight lines can be flown in any direction, the orientation of the images
relative to cardinal directions can vary; this is unlike most map products, which tend
to place north towards the top of the map. The overlap of flight lines is required to
ensure complete coverage and to facilitate the use of photogrammetric methods
for creating maps from the imagery. Complete coverage of an area is addressed by
having flight lines imaged with approximately 30% sidelap. Along any given flight
line, overlap between adjacent photographs is typically 60%. Although this increases
the cost of obtaining full coverage of an area, imaging the same portion of the Earth's
surface in adjacent photos by this amount creates *stereopairs* (Figure 9.5).

FIGURE 9.4

Photographic overlap
along flight lines and
sidelap between adjacent
flight lines.
Source: Elisabeth Nelson

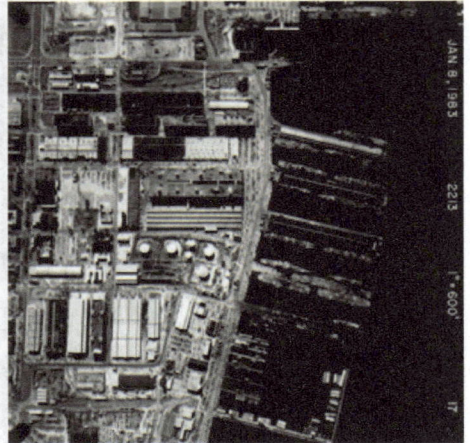

FIGURE 9.5
Stereopair of Norfolk, Virginia, 1983. Photos: *Department of Geography, UNCG.*

Stereopairs show the same portion of the Earth's surface, but from slightly different camera positions. When viewed together, using the appropriate photogrammetric equipment, the images are perceived as a 3D model, or *stereomodel*, of the area. The principle behind this concept rests on our own eye-brain system. Each of our eyes views its surroundings from a slightly different position; the brain takes the information gathered from each eye and merges it, allowing us to generate depth perception and view our world in 3D.

GEOMETRIC DISTORTION

The geometric distortion inherent in unprocessed aerial photos takes several forms. Some of the most important include height (scale) distortion, tilt displacement and relief (topographic) displacement. *Height distortion* occurs because the height, or elevation, of the terrain varies as the plane flies along a flight line. Objects at higher elevations are closer to the camera and will appear larger in the photograph than

FIGURE 9.6
Height Distortion: An object at a higher elevation will appear larger than an identical object at a lower elevation.
Source: Elisabeth Nelson

similar objects at lower elevations, which means that scale varies over the photograph (Figure 9.6).

When the optical axis of the camera is not precisely vertical during image acquisition, the resulting photograph will also show effects of *tilt displacement*. Almost all vertical photographs will have at least a small degree of tilt due to a variety of factors, including wind and aircraft maneuvers like banking and climbing. What is interesting about tilt displacement is that there is an order to the severity of the distortion. Tilt displacement affects the shapes of objects in the image, with object distortions increasing radially inward or outward from the *isocenter*. The isocenter, on a true vertical photograph, would be the same as the principal point and nadir. Since all vertical photographs have some degree of tilt distortion, though, a better definition would be the point on the photograph that bisects a line connecting the principal point and nadir (Jensen, 2000). Because this type of distortion has a regular pattern, computer algorithms can be used to remove it from an image.

With *relief* or *topographic displacement*, the tops of objects in an image are displaced relative to their bases. There is also a radial pattern to this type of distortion, with objects located above the terrain leaning away from the nadir of the photograph (Figure 9.7).

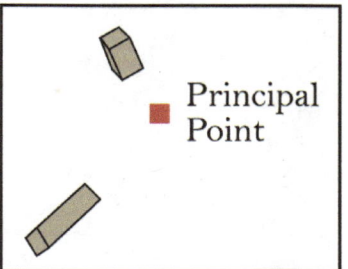

Distortion, again, increases towards the edges of the photograph. This type of displacement is caused by a combination of the perspective geometry of the camera lens and the varying elevation of the terrain in the image. While it may be processed out of the image to create a product that can be used as a base for mapping, relief displacement can also be useful to the photointerpreter, as it is harnessed for both stereoviewing and for calculating heights of objects in the image.

Basic Photogrammetric Techniques

Photogrammetric techniques range from methods for determining distances, areas, and heights of objects on an image to using the image to generate digital elevation models, orthophotos, and other related products. Below are a few examples of some of the simpler techniques involved in this science, along with a brief overview of some of the products that can be derived from aerial photography.

WORKING WITH SCALE ON AN AERIAL PHOTOGRAPH

As with maps, scale is a central feature of the aerial photograph. By defining the size relationship between a real-world feature and its corresponding image on the aerial photo, photogrammetrists can use it to both measure features and plot distances. The scale of a photo is formally defined as the ratio of the camera lens focal length (f) to the flight altitude (H) of the plane, written mathematically as: $\frac{f}{H}$. Although this ratio looks nothing like the standard $\frac{map\ distance}{ground\ distance}$ ratio we are used to working with when handling maps, some simple geometry proves it is a surrogate form (Jensen, 2000). As

Figure 9.8 shows, line **XY** on the ground is imaged as line **xy** on the photograph. The ratio $\frac{xy}{XY}$, then, is equivalent to $\frac{Photo\ Distance}{Ground\ Distance}$. From similar triangles **ZXY** and **zxy**, then, $\frac{f}{H}$ is also equivalent to $\frac{Photo\ Distance}{Ground\ Distance}$. Remember, however, this scale is not consistent across the image. Much like a projected map's principal scale is only true at the point or line of tangency, the photo's scale is only true at the principal point, and may vary somewhat across the image due to changes in object heights and terrain elevations.

The focal length and flight altitude are usually stamped at the top of the image. To compute the ratio, both parameters must first be transformed into the same units of measurement. For example, on a photograph with a 6 inch focal length and flight altitude of 5,000 feet, the steps are to convert both measurements to inches, calculate the ratio, then convert the ratio to representative fraction form by dividing out the numerator to arrive at 1 for the upper part of the ratio:

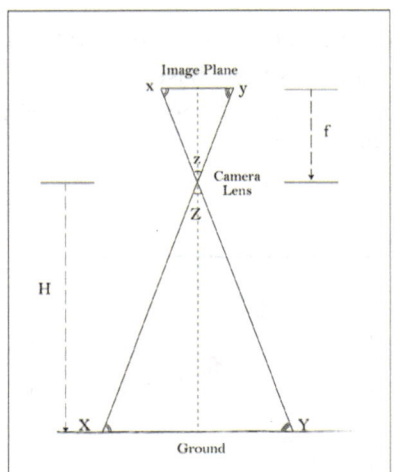

6 inch focal length/ 5,000 ft. (12 inch) flight altitude =

$$\frac{6}{60,000} = \frac{\frac{6}{6}}{\frac{60,000}{6}} = \frac{1}{10,000}\ or\ 1{:}10{,}000$$

If the image is not annotated with focal length and flying altitude, another way to estimate its scale would be to measure the distance on the image between two features, then to compare that distance to one computed from a map that has a defined scale. Two formulas are useful here (Avery and Berlin, 1992). For maps, this formula would be: *Ground Distance = Map Distance x Map Scale Factor*, where the *Map Scale Factor* is defined to be the inverse of the map scale. The corresponding formula for an air photo would be *Ground Distance = Photo Distance x Photo Scale Factor*. Since both formulas represent ground distance, it follows that:

Map Distance x Map Scale Factor = Photo Distance x Photo Scale Factor

Suppose, then that the distance between **A** and **B** on an image with an undefined scale is 0.2 inches. On a map of the same area, with a scale of 1:10,560, the same distance between those features measures 0.1 inch. If we input the 'knowns' of our problem into the formula above, we have one unknown we can solve for: *Photo Scale Factor*.

$$0.1\ inch\ x\ 10{,}560\ inch = 0.2\ inch\ x\ Photo\ Scale\ Factor$$
$$\frac{1{,}056}{0.2} = \frac{0.2}{0.2}\ x\ Photo\ Scale\ Factor$$
$$5{,}280 = Photo\ Scale\ Factor$$

Since Photo Scale Factor is the inverse of Photo Scale, if we invert the answer, we will have the photo's estimated scale.

$$1{:}5{,}280 = Photo\ Scale$$

DETERMINING OBJECT HEIGHTS

Aerial photographs are 2D representations of a 3D world. In photographing that environment, all the 3D attributes of features, such as object heights, are forced onto a plane with no vertical dimension. Unlike a *planimetric* map, however, in which the features are all plotted as if they were located at the same elevation, those same feature locations on an air photo are influenced by many factors, including the local scale at that point, the feature elevation and its distance from the nadir. These relationships result in a distinct topographic displacement pattern, where any feature above or below the mean ground level of the photo is displaced from its true horizontal location, and the tops and bases of many features can be easily discerned. This displacement can be harnessed to estimate the heights of objects in the image by using the following formula (Avery and Berlin, 1992):

$$\text{Object Height} = d/r \, (H)$$

where: d = length of the object from top to bottom
r = distance from top of object to principal point of image
H = flying height of aircraft during image acquisition

If, as Figure 9.9 shows, the length (d) of the object being measured was 0.10 inches, the top of the object was 0.40 inches from the principal point (r), and the flying height (H) was 2,000 feet, then the object height would be estimated as:

$$0.10 \text{ inches}/0.40 \text{ inches} \, (2000 \text{ ft.}) = 500 \text{ feet}$$

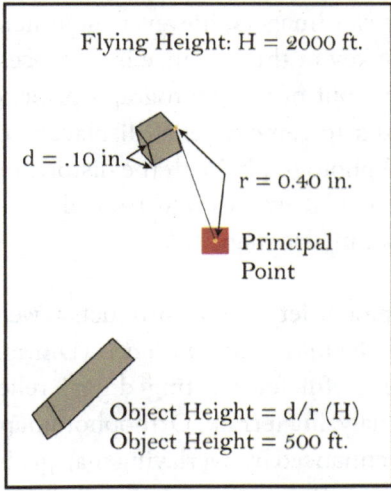

FIGURE 9.9
Estimating the heights of objects using topographic displacement.
Source: Elisabeth Nelson

Flying Height: H = 2000 ft.

d = .10 in.

r = 0.40 in.

Principal Point

Object Height = d/r (H)
Object Height = 500 ft.

UNDERSTANDING STEREOSCOPY

Another area in which the height or relief displacement of objects on a vertical photograph can be used to an advantage is when working with pairs of adjacent aerial photographs. Because such images are planned with a 60% overlap, the images can be aligned in such a way that each eye can essentially view the same scene, with a slight distance shift occurring in feature positions from one image to the next. This shift, or *parallax displacement*, is converted by the eye-brain system into a 3D model of the area. Experts often view such images with their unaided eyes and perceive depth; most others, however, will find that using a *stereoscope* in conjunction with the aligned images to be the most efficient means of creating the necessary depth perception.

The stereoscope is a photogrammetric instrument that enhances one's ability to fuse visually two images of an area with slightly different perspectives into a *stereomodel* or 3D environment. It works by presenting one image to the left eye and the other, adjacent image to the right eye. The parallax displacement between the two images is used by the eye-brain system to perceive the images in 3D form. The geometry of the resulting stereomodel then allows scientists to make precise 3D measurements. Over time, many techniques and instruments have been developed to enhance this basic process. One example particularly important to topographic mapping involves the use of a *stereoplotter*.

Stereoplotters are instruments that allow the operator not only to view the environment in 3D, but also to trace paths of constant elevation, creating the contours we see on our topographic map series. This type of instrument is designed to control both the horizontal and vertical movements of a measuring mark. By setting the height of the measuring mark to rest on the terrain surface, the operator can trace contour lines for that particular elevation and store them digitally. The topographic maps produced today follow this procedure, not only for creating the contours but also for identifying and measuring all other features. The operators even add attributes to each feature as they work, creating both a topographic map as well as additional vector GIS data that can be used in GIS analyses (Aronoff, 2005).

CORRECTING GEOMETRIC DISTORTION

While creating stereomodels and calculating relief displacements are crucial photogrammetric analyses, there are also other ways those in GIScience might use aerial photographs. One example would be to pair or overlay a photograph with vector GIS data. Because of the distortions inherent in the raw imagery, however, alignment issues with these datasets will cause problems. The key in this instance is to process relief displacement and other geometric distortions out of the photograph prior to using it. An *orthophotoscope* is the instrument used to remove relief displacement and produce an *orthophoto* from the original aerial photograph. With the distortions removed, every point in the orthophoto appears as if it was viewed from directly above, producing a photo that is equivalent to a planimetric map.

Today's orthophoto technology is capable of producing a variety of other products as well, including digital arrays of terrain elevations called *Digital Elevation Models* (DEMs), *orthophotomaps* and *orthophotoquads*. DEMs are useful for creating digital relief maps, 3D representations of an environment, and analyzing terrain. Orthophotomaps are orthophotos in quadrangle form that have been enhanced by overlaying map grids, placenames, symbols, and contours. Black and white orthophotoquads are similar to our 1:24,000 topographic series, but typically have less information than the standard quadrangle; these tend to be used in specific applications, such as site selections or tax assessments (Monmonier and Schnell, 1984).

Landscape Interpretation

In addition to the information derived from aerial photographs using photogrammetric techniques, photointerpreters also use a variety of visual indicators to identify objects and infer information about a region. The traditional framework from which they work makes use of eight visual elements: shape, size, tone, texture, shadow, site, association, and pattern (Aronoff, 2005). What follows are some simple examples to introduce you to these elements and how they are used in landscape interpretation. Today's applications of these elements in specific fields, such as archaeology, forestry, agriculture and urban landscapes, will quickly require expertise beyond these examples, the rubrics of which are outside the scope of this text.

As we saw in our discussion of basic visual variables used in mapping, shape is a property of objects to which we are particularly sensitive visually. Unique shapes, such as athletic fields, or the building footprint of a structure like the Pentagon, are easily identified using this element alone if you know the characteristics for which you are looking (Figure 9.10).

FIGURE 9.10
Identifying features using the visual element of shape: Pentagon (left), William-Brice football stadium, Columbia S.C. (right). Photos: *U.S. Geological Survey, Department of the Interior.*

Absent the unique distinctiveness, you can typically designate an object as at least of cultural or natural origins. Objects with regular, geometric shapes and straight lines (roads, for example) are usually the result of human construction, making them a cultural object. Naturally-occurring features like rivers, on the other hand, are more likely to be irregular in shape and non-geometric. With a little more expertise or additional information, you can even distinguish between objects that have similar shapes but lack unique distinctive features. One example here would be identifying an object as a road versus a railroad (Figure 9.11). While these features tend to look similar on many photographs, a closer look will reveal that railroads are often more linear, have gentler curves than roads, and lack perpendicular intersections (Rabenhorst and McDermott, 1989).

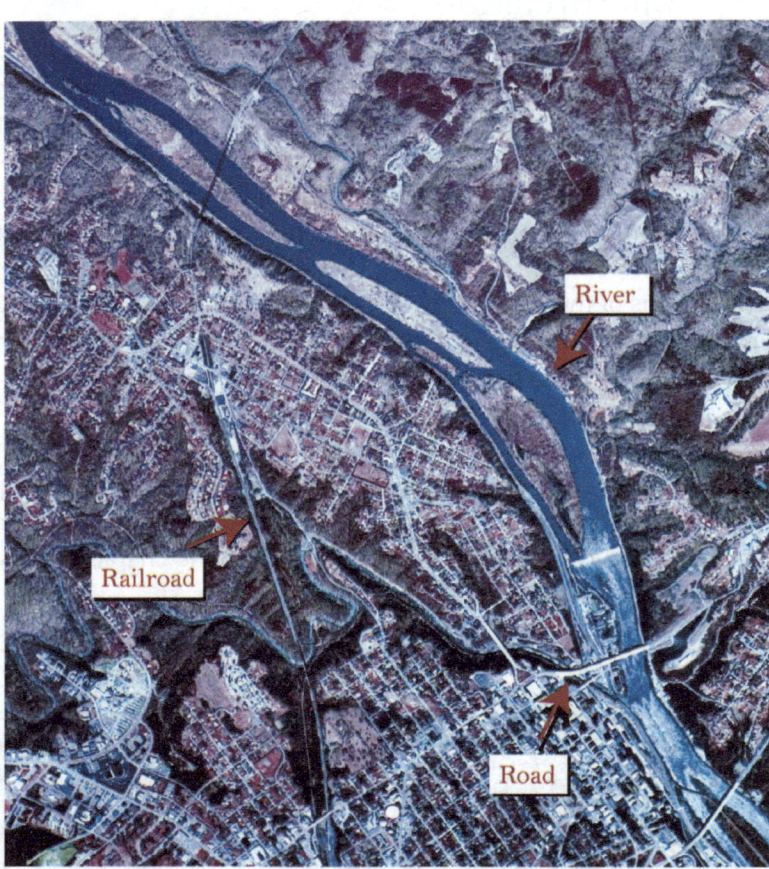

FIGURE 9.11
Identifying features using the visual element of shape: roads, railroads, and rivers, Lynchburg, V.A. Photo: *U.S. Geological Survey, Department of the Interior.*

Another instance in which shape may be useful even when it isn't unique is when it is considered in the context of surrounding land use. Take as an example a public school in the U.S. The building shape itself will look similar to many other types of buildings, but playing fields, parking lots, and residential areas will surround it. Each of these, when considered in conjunction with the building, helps to identify it as a school, as opposed to a mall or a single-family residence (Figure 9.12). *Association* is the interpretive element that describes this process of using the functional connections of surrounding features to help identify an object.

Size, the measure of an object's dimensions, is another fundamental visual variable of which we make use in landscape interpretation. Object sizes can be used in both an absolute sense and a relative sense to help identify a feature. Absolute sizes are calculated photogrammetrically, and are often used to classify a feature as belonging to a particular group. Houses, for example, could be categorized by absolute size. Doing this then allows the photointerpreter to use house sizes to infer housing values (Avery and Berlin, 1985). Absolute sizes of known features also help us to appreciate the sizes of adjacent, sometimes unknown, features. This use of relative size refers to the process of comparing an object's size to those nearby. Jensen (2000) provides an interesting example of this, where the average lengths of trailers on tractor-trailer rigs (45-50 feet) can be used to gauge how large adjacent warehouse buildings are on an image.

FIGURE 9.12
Identifying features
using the visual element
of association: schools
versus malls, Greens-
boro, N.C. Photos: *U.S.
Geological Survey, De-
partment of the Interior.*

Aerial photographs also provide access to *shadow* information if an object has height, and to *tone or value* information. Shadows are important because they sometimes expose details that are difficult to see from directly overhead. The shadows of steeples, for example, are often used in identifying buildings as churches. Tone or value refers to how bright an object is or to its dominant color. Tone in a traditional aerial photograph is created through a chemical reaction during the development of the camera's film. This film is light-sensitive; the camera records the amount of light energy reflected from the portion of Earth being imaged, and when the film is developed, variations in light intensity shows up as differences in tone.

This interpretive element is somewhat trickier than the previous, because an object's tone is dependent upon 1) how much light it reflects and how much of that light is captured by the camera as it records the image, and 2) its texture. If an object reflects a lot of light, then it will appear lighter on the resulting image. Smoother surfaces, such as roads and bare fields, generally tend to have higher reflectance values and show up on an image as lighter tonal values. Rougher textures, on the other hand, scatter light so that less is reflected back to the camera, resulting in darker tones on the image. Areas of forest are a good example here (Figure 9.13); leaves and branches intermix and create shadows so that when imaged, the area appears coarser and darker (Aronoff, 2005). *Texture* is considered an interpretative element in and of itself; it is used frequently to distinguish types of agriculture, where some crops appear coarser than others do. Corn, for example, generally has a coarser texture than wheat (Rabenhorst and McDermott, 1989).

The last two interpretative elements to consider are *pattern* and *site*. When objects are arranged in a repetitive way, the result is a design called a pattern. Trees that are planted in rows create a pattern we associate with orchards. Tombstones in a cemetery also create a distinctive pattern that can be recognized on aerial photographs. Another interesting example of pattern in the United States is the presence of long lots along the Mississippi River (Figure 9.14). While much of the United States was surveyed using the Public Land Survey system, this was not the case along the Mississippi River, which had already been surveyed by the French during the time they held the land. The French survey created long, narrow parcels of land that gave each owner access to the Mississippi, as it was the primary means of travel during that time. This pattern also ensured each owner had similar areas of land quality, which tended to vary as distance from the river increased (Rabenhorst and McDermott, 1989). While both the French long lot and the Public Land Survey system produced rectangular patterns, the character of the French long lots – their orientations and the ratios of the sides of the rectangles – make them distinct on imagery. The French long lot in the U.S. is also an example of a *site*, which is an interpretive element that has a unique physical or socioeconomic characteristic. In this case, the French survey created a land-tenure system unique to the Mississippi River.

FIGURE 9.13
Identifying features using the visual elements of tone and texture: Forests, roads, and fields, Reidsville, N.C. Photo: *U.S. Geological Survey, Department of the Interior.*

While these eight elements have been discussed largely in isolation, in practice they most often are considered simultaneously during interpretation. Whether alone or in combination, however, they can be used to complete tasks that produce derived information from the raw photograph. These tasks include classifying the imagery, counting objects visible on an image, measuring objects and delineating regions with homogeneous tones and textures (Aronoff, 2005).

FIGURE 9.14
Identifying features using the visual elements of pattern and site: French long lots along the Mississippi River, St. James Parish, Louisiana. Photo: *U.S. Geological Survey, Department of the Interior.*

References

Aronoff, Stan. 2005. *Remote Sensing for GIS Managers.* Esri Press: Redlands, California.

Avery, T.E. and G.L. Berlin, 1992. *Fundamentals of Remote Sensing and Air Photo Interpretation.* 5th ed. Macmillan Publishing Company: New York, New York.

Campbell, J.C. 1984. *Introductory Cartography.* Prentice-Hall, Inc.: Englewood Cliffs, New Jersey.

Campbell, J.B. 1996. *Introduction to Remote Sensing.* Guilford Press: New York, New York.

Dickinson, G.C. 1979. *Maps and Air Photographs.* 2nd ed. Edward Arnold, Ltd.: London, England.

Gray, Carroll. 2002. *The First Five Flights: The Slope and Winds of Big Kill Devil Hill – The First Flight Reconsidered.* Retrieved from http://www.thewrightbrothers. org/fivefirstflights.html Last accessed 28 Dec 2012.

Jensen, J. R. 2000. *Remote Sensing of the Environment: An Earth Resource Perspective.* Prentice-Hall, Inc.: Upper Saddle River, New Jersey.

Lillesand, T., Keifer, R.W., and J. Chipman. 2007. *Remote Sensing and Image Interpretation.* 6th ed. John Wiley & Sons: New York, New York.

Monmonier, M. and G.A. Schnell, 1984. *Map Appreciation.* Prentice-Hall, Inc.: Englewood Cliffs, New Jersey.

Rabenshorst, T.D. and P.D. McDermott, 1989. *Applied Cartography: Introduction to Remote Sensing.* Merrill Publishing Company: Columbus, Ohio.

Thrower, N.J.N. 1996. *Maps and Civilization: Cartography in Culture and Society.* University of Chicago Press: Chicago, Illinois.

Exercise 9

Concepts in Photogrammetry and Landscape Interpretation

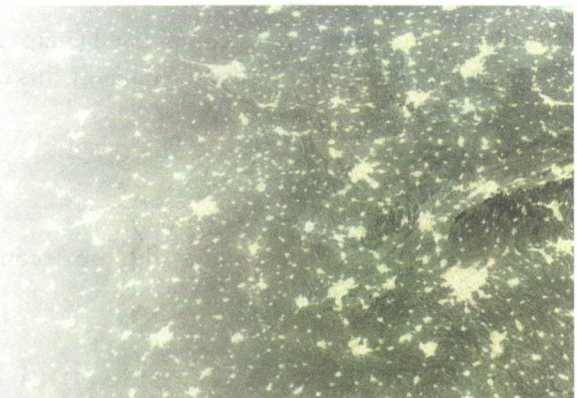

Exercise 9 provides experience with some of the basic photogrammetric and landscape interpretation concepts used in aerial photography.

OBJECTIVES

- Use an image's fiducial marks to determine its principal point
- Estimate the scale of a photo
  - Using flying height and camera lens focal length
  - Using known features
  - Using the distance between two features and a map
- Estimate the height of an object using flying height and distance from the photo's principal point
- Use visual elements to identify and infer information about landscape features

SOFTWARE INFORMATION

The questions for Exercise 9 require Adobe® Acrobat® (Reader DC) and Google Earth. Adobe Acrobat (Reader DC) is available in many geospatial labs (or download for free at https://get.adobe.com/reader/). Additionally, Google Earth can be downloaded for free at http://www.google.com/earth/

DATA

The data for the exercises is available from the Kendall Hunt Student Ancillary site. See the inside front cover for access information. You may also be directed to download the data from a different location by your instructor.

1. Download the data as instructed by your instructor
2. Save the file to a location where you have read/write privileges (USB key, home computer, class server)

The file you just saved is in a compressed (*.zip) format, and was created using **7-zip** freeware (http://www.7- zip.org). To use the data for this exercise, you must decompress it using the same software.

3. Locate the zipped file that you saved
4. Right-click the file
5. Choose *7-Zip - Extract Files*
6. Click *OK* to create a folder with the decompressed data

Fiducial Marks and the Principal Point

Fiducial marks are images of markers built into the aerial camera. When a photo is taken, these marks are reproduced on the sides or in the corners of the photograph. Connecting opposite pairs of fiducial marks with perpendicular lines will determine the photo's principal point. The principal point is the geometric center of the photograph.

1. Navigate to the location where you saved the *Chapter09* folder containing the data for this exercise
2. Double-click the folder to see its contents
3. Right-click *Aerial_Photography_1* and *Open with – Adobe Reader DC*

QUESTION 9.1

Aerial Photograph 1 (Figure 9.1) is a photo of the Norfolk, VA area. Norfolk, a coastal city in the state of Virginia, is known for its transportation and military prowess. The largest Navy base in the world is located here, as is the corporate headquarters of Norfolk Southern Railway.

Using the photograph, connect the fiducial marks to locate the principal point of the photograph – mark this location on the photo.

What Cultural landscape feature is located at this photo's principal point?

a. Airport
b. Storage Tanks
c. Warehouses
d. Ships

FIGURE 9.1

Estimating Photo Scale

You can derive the scale of an aerial photograph in several ways: as the ratio of camera lens focal length (f) to flying height (H) by comparing the distance between two objects on both the photo and a map with a scale, and by using an object on the photo with known length.

QUESTION 9.2

You have been given a folder of imagery with no annotation; however, the folder does contain a sheet of paper with the following information:

COMBAT --- Commercial Block Analytical Triangulation

Project: University of North Carolina Greensboro; Strip 3, 3 Photos

Photo Scale:

Flight Altitude (AMT): 8,325 feet

Camera Focal Length: 6 inches

1

Based on the Flight Altitude and the Camera Focal Length, compute the scale for this photo, stating it as a representative fraction for your final answer. What is the scale of the photograph, stated as a representative fraction?

a. 6:8,325
b. 1:1,388
c. 6:99,900
d. 1:16,650

QUESTION 9.3

Aerial Photograph 1, of Norfolk, VA as previously shown in Figure 9.1 is a 30-year-old image you found in the back of a filing cabinet. You don't know who took the image, what kind of camera they used, or the altitude of the plane.

You need to estimate the scale of the image. You believe that you can work with an aircraft carrier in the upper left quadrant of the photo to acquire the scale of the image. Aircraft carriers are easy to identify because they have angled flight decks for landing planes (Figure 9.2).

The number on the bow (facing toward land) identifies the carrier. A little detective work via Google turns up a list of all U.S. Navy aircraft carriers by number. CV-62 is the *Independence*, which is listed as 1,070 feet in length.

FIGURE 9.2

Use *Aerial_Photograph_1* to measure the image length of the ship.

1. You can print *Aerial_Photograph_1* or you can work in Adobe Acrobat (Reader DC)
2. If you are working in Adobe Acrobat Reader DC, go to the *Tools* menu and type *Measure* in the Search (Find your tools here). Click *Open*. Your *Measuring Tool* should appear in the toolbar above the photo.
3. Zoom in on the aircraft carrier, and click on your *Measuring Tool* to estimate it's length in the photo.

Based on your measurement from the photo and the ship's real-world length, compute the scale for this photo, stating it as a representative fraction for your final answer (be sure to show your work).

What is the scale of this map, stated as a representative fraction? _____

Aerial Photograph 2, of the University of North Carolina Greensboro campus in Greensboro, NC (Figure 9.3A) is a section of a digital aerial photograph. The scale of the photograph was not included in the metadata, but you have access to a 1:14,500 scale reference map of the area (Figure 9.3B) which you can use to estimate the photo's scale.

1. Right-click *Aerial_Photo_2_Map.pdf* and choose *Open with – Adobe Reader DC*

Use *Aerial_Photo_2_Map.pdf* to measure the distances between points A and B on the photo and the map.

2. You can print *Aerial_Photograph_1* or you can work in Adobe Reader DC
3. If you are working in Adobe Acrobat Reader DC, go to the *Tools* menu and type *Measure* in the Search (Find your tools here). Click *Open*. Your *Measuring Tool* should appear in the toolbar above the photo.

4. On the photo, click at A, then move to B and double-click to finish the measurement
5. Repeat the process on the map

FIGURE 9.3A

FIGURE 9.3B

EXPLORING GEOSPATIAL TECHNOLOGY

QUESTION 9.4

Measure the distance between the labeled A and B points on **both** the photo and the map.

How far is the distance in inches between A and B on the photo? _____

How far is the distance in inches between A and B on the map? _____

QUESTION 9.5

Using the map's scale, determine the **real-world** length of road segment on the map that lies between these points.

Record the length in feet: _____

QUESTION 9.6

Using the results obtained in Question 9.5 and the photo distance between the same two points (A and B), calculate the scale of the photo. State your answer as a representative fraction (be sure to show all work).

Representative Fraction: _____

Estimating an Object's Height

Features above or below the mean ground level of a photo are displaced from their true horizontal location, and often both the tips and bases of some objects can be seen. If this is the case, then the displacement can be used to estimate the height of an object if you also know the flying height of the plane and can locate the principal point of the photo:

Object Height = d/r (H)

Where:

d = length of the object from top to bottom
r = distance from top of the object to the principal point of the image
H = flying height of the aircraft during image acquisition

QUESTION 9.7

Aerial Photograph 3 (Figure 9.4) of the Appalachian State University campus in Boone, NC shows a building just north of the football field (A).

    a.   Find the principal point of the photograph and mark the location on the photo

    b.   If the flying height of the plane that took this photograph was 1,500 feet, approximate how tall is the building Labeled A? (Be sure to show your work)

Answer: _____

FIGURE 9.4

Identifying Landscape Features

In addition to the information derived from aerial photographs using photogrammetric techniques, photo interpreters also use a variety of visual indicators to identify objects and infer information about a region. The traditional framework from which they work makes use of eight visual elements: shape, size, tone, texture, shadow, site, association, and pattern.

QUESTION 9.8

Identify the features in the following aerial photos (Figures 9.5a-d).

FIGURE 9.5

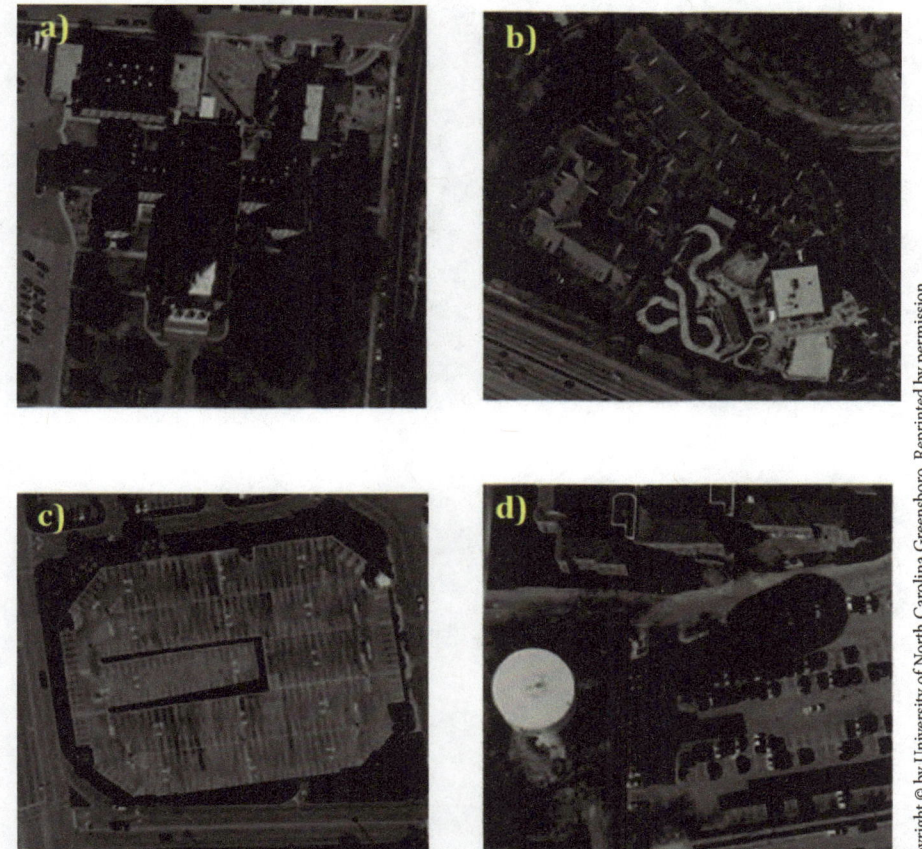

a) Top Left	b) Top Right	c) Bottom Left	d) Bottom Right
a. Mall	a. Racetrack	a. Parking Deck	a. Grain Silo
b. School	b. Water Park	b. Shopping Center	b. Lighthouse
c. Church	c. School	c. Tennis Courts	c. Cell Tower
d. Hospital	d. Farm	d. Football Stadium	d. Water Tower

Chapter 10

Remote Sensing

Traditional aerial photography is but one method available today for acquiring imagery of the Earth. By the end of World War II, the demands of the military for more and better cameras, films, and platforms for capturing images of our landscapes had created another technological boom in the United States. With an influx of German scientists leading the way, the U.S. Space Program gained momentum, and the rocket became the new platform for this growing geospatial technology (Thrower, 1996). Rocket technology, along with the development of artificial satellites and non-photographic scanning systems, prompted the director of the U.S. Office of Naval Research, *Evelyn Pruitt*, to coin a new term for this expanding field: *remote sensing*. She considered this new technology, based on electronics (scanners) and radiation physics (the entire electromagnetic spectrum), a separate field from traditional aerial photography, the foundations of which are chemistry (film) and optics (cameras). Today, however, the term *remote sensing* covers both, and is typically defined as the process of acquiring information about Earth using any technology that is not in direct contact with the Earth's surface (Monmonier and Schnell, 1984).

Electromagnetic Radiation

All remote sensing technology, whether based on traditional aerial photographic techniques or the non-photographic technology of satellites and electronic scanner systems, is designed to sense light energy that is either reflected or emitted from ground-based objects. The principal type of light energy detected is *electromagnetic radiation,* released from the Sun. This radiation is composed of vibrating electric and magnetic fields that travel through space in waves. A stream of photons, or discrete bundles of light energy, comprises each wave, which moves at the speed of light. Different types of radiation, such as visible light or radio waves, consist of photons with different energy levels, wavelengths, and wavelength frequencies (NASA, 2012).

Of these radiation characteristics, two are key for understanding how scientists obtain information from remote sensing data: *wavelength* and *frequency.* A wavelength is defined as the distance from one wave peak to the next; a wave's frequency is the number of these peaks that pass a fixed point in a given timeframe. As Figure 10.1 shows, shorter wavelengths have higher frequencies and greater amounts of energy than longer wavelengths. This relationship can be modeled mathematically as: $c = f\lambda$

Where c is the speed of light (3×10^8 m/s), f is the frequency of the wavelength (cycles per second, Hz), and λ is the wavelength (m).

So that the product of f and λ is always equal to the speed of light, the frequency of the wave must always be inversely related to its wavelength.

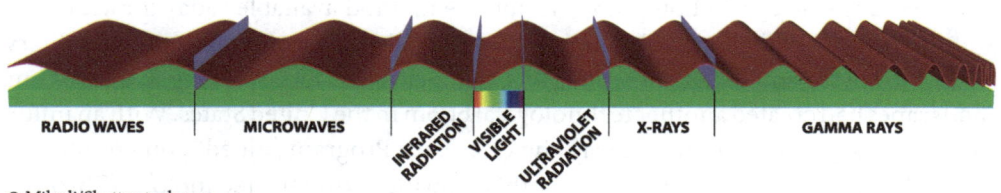

ELECTROMAGNETIC SPECTRUM

RADIO WAVES   MICROWAVES   INFRARED RADIATION   VISIBLE LIGHT   ULTRAVIOLET RADIATION   X-RAYS   GAMMA RAYS

© Milagli/Shutterstock.com

FIGURE 10.1
The electromagnetic spectrum: wavelength and frequency.

Scientists categorize the continuum of electromagnetic radiation by frequency to apply these concepts and examine how they influence information extraction (Natural Resources Canada, 2008). Once categorized in this manner, the result is referred to as the *electromagnetic spectrum (EMS).*

Atmospheric Interactions

Before the energy of the electromagnetic spectrum can be detected by remote sensing systems, it must first make it through our atmosphere and then back to the sensor. This is no mean feat, as our atmosphere contains a variety of particles and gases that either absorb, scatter, or refract many of these wavelengths before they ever reach the surface of Earth. One of the interesting things about this interference is that it is *wavelength selective;* as Figure 10.2 shows, some wavelengths pass through the atmosphere untouched, while others are restricted in some manner (Campbell, 1996; Avery and Berlin, 1992).

FIGURE 10.2
Atmospheric
interactions: absorption
bands and atmospheric
windows. *Courtesy
NASA/JPL-Caltech.*

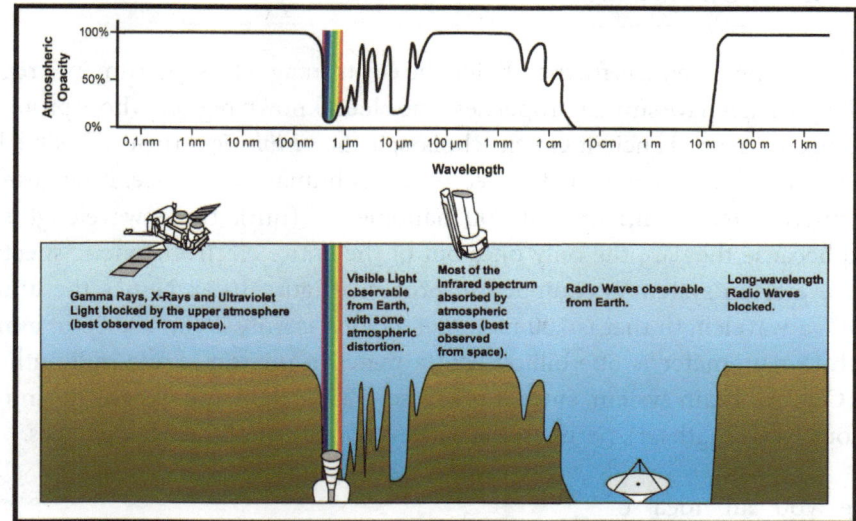

Restricted wavelengths aren't as useful for remote sensing applications, as the amount of energy reaching the ground is reduced, leaving less to be detected when it is reflected off ground-based objects. These areas of the EMS are called *absorption bands,* because the wavelengths are absorbed by such atmospheric gases as carbon dioxide, water vapor and ozone (NASA, 2012). Ozone, for example, absorbs portions of the ultraviolet wavelengths, while carbon dioxide absorbs portions of mid-infrared and far-infrared wavelengths and water vapor absorbs portions of near-infrared and thermal infrared wavelengths.

These same atmospheric gases also play a role in *scattering* the shorter wavelengths of the EMS, including the visible blue wavelengths. As these wavelengths enter our atmosphere, they are absorbed, then redirected or *scattered*, away from their initial direction of travel by particles that are much smaller than the wavelengths. Known as *Rayleigh scattering*, these wavelengths are diffused throughout the sky, which is why our sky appears blue. Rayleigh scattering is heavily wavelength dependent; it is inversely proportional to the 4th power of the wavelength, so shorter wavelengths are much more affected than longer ones (Nave, 2012). This type of scattering also reduces image contrast.

Those wavelengths that are not restricted, on the other hand, form *atmospheric windows*. These regions are important for remote sensing because they are the most suitable regions of the spectrum around which to develop remote sensing systems. They include the visible wavelengths and near-infrared wavelengths, as well as the thermal infrared, microwave, and to some extent the shorter ultraviolet wavelengths (Avery and Berlin, 1992).

The Visible Spectrum

In remote sensing, scientists further divide the electromagnetic spectrum into regions of wavelengths that have similar properties. The oldest known region, whose properties were studied by several ancient Greek scholars, is the *visible light* region. Visible light is that region of the electromagnetic spectrum that humans use to see. It ranges from 0.4-0.7 micrometers (μm,) or 400–700 nanometers (nm); these wavelengths are grouped because they are the only ones out of the entire electromagnetic spectrum that our eye-brain system is equipped to process (Figure 10.3). Notice the units of measure – a wavelength that is 700 nanometers long is only 7/10,000,000 of a meter in length (a nanometer is one billionth of a meter; a micron is one millionth of a meter)! Our eye-brain system, since it processes visible light and assigns meaning to the various wavelengths, is, perhaps, the earliest of our remote sensing devices.

To give you an idea of how our eye-brain system functions in such a manner, consider the following simplified description of the process. Imagine yourself walking through a wooded area on a sunny, summer day. The sunlight shining down on you includes all the wavelengths of the visible spectrum in equal proportions. When this light shines on an object in your view, like the leaves on

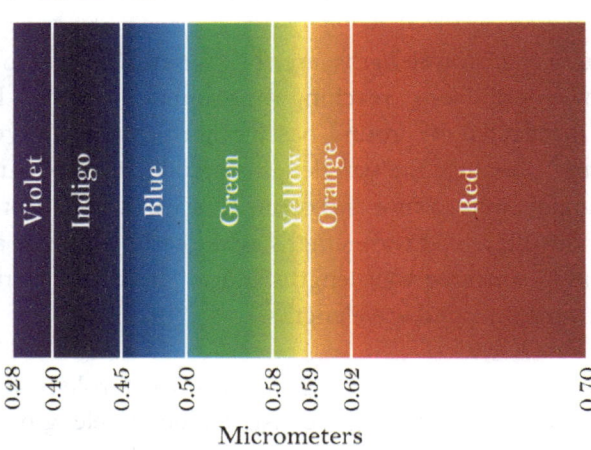

FIGURE 10.3
Visible spectrum.
Source: Elisabeth Nelson

a tree, the chemical makeup of that object will *absorb* some of those wavelengths, and *transmit* or *reflect* others. In the case of the summer leaves, the shorter 04.-0.5 μm wavelengths and the longer 0.6-0.7 μm wavelengths are absorbed by the leaves' chlorophyll, while those in the 0.5-0.6 μm range are reflected and processed by your eye-brain system as you look at the trees (Monmonier and Schnell, 1984).

What is particularly unique about our personal remote sensing system is the meaning the brain assigns to each of the wavelengths. As our system processes reflected light, it associates a hue, or dominant color to each wavelength. In this case, that hue would be green. If an object reflected primarily the 0.4 – 0.5 μm range, then we would perceive it as being blue; if it reflected primarily in the 0.6-0.7 μm range, we would see it as red. Each of these narrower wavelength frequencies is known as a *spectral band*.

USING VISIBLE SPECTRAL BANDS IN REMOTE SENSING

Because all objects absorb or transmit some wavelengths of the electromagnetic spectrum, while reflecting others, each has a unique pattern of reflected radiation, known as a *spectral signature*. These patterns form the basis for remote sensing, and they can be graphed using a *spectral signature graph* (Figure 10.4). This graph creates the signature, or response curve, of an object by plotting the wavelengths of light along the x-axis of the graph and the percentage of that light that is reflected from each band on the y-axis.

FIGURE 10.4
Spectral signature graph for lawn grass. *U.S. Geological Survey, Department of the Interior.*

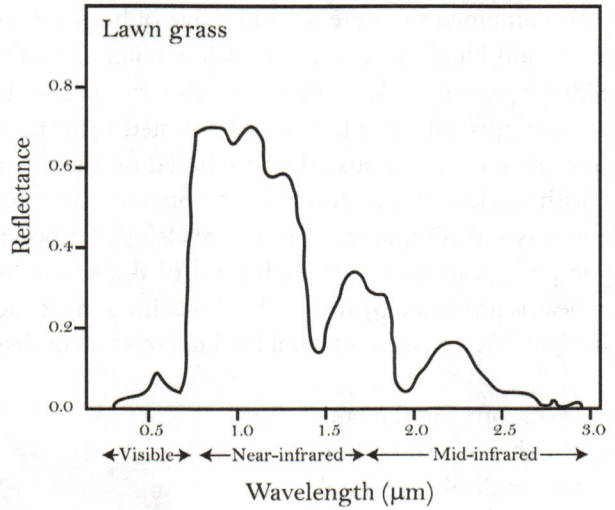

The percentage (*p*) of reflected light is determined in a two-step process. First, the total amount of energy, or *incident energy (I)*, striking an object is calculated by measuring and summing the amounts of energy *reflected (R)*, *absorbed (A)*, and *transmitted (T)* per wavelength for that object: $I = R + A + T$

Then, the amount of reflected energy is divided by the incident energy and the resulting fraction is multiplied by 100 (Shellito, 2012): $p = (R/I) * 100$

In the visible spectrum, for example, water, soil and vegetation all form distinct spectral signatures that allow us to distinguish them on photographs or digital images.

Each of these bands has intended principal applications that help guide our use of them in remote sensing applications. The visible blue band, for example, is the best of the three primary visible bands (red, green, blue) for penetrating water, so examining the landscape through this particular lens may allow us to assess water quality or map coastal zones. The visible green band is also used in water quality studies, as well as for studying urban infrastructures. Remote sensors use the visible red band in vegetation studies and for determining soil and geologic boundaries. All three bands are useful for identifying cultural features (Aronoff, 2005).

DISPLAYING VISIBLE SPECTRAL BAND DATA

The hues that we associate with the wavelengths of the visible spectrum are not actual properties of light, but rather our eye-brain system's response to light. We are able to process light in the visible spectrum because our eyes are equipped with three types of color receptors or *cone cells* that are each sensitive to the red, green, and blue spectral bands of the visible spectrum. Through a sophisticated processing network, input from each of these cones is combined to create the full range of hues we associate with this spectrum. Red, green, and blue, then, when modeling color from the perspective of light, are called *additive primary colors*. They are primary because by varying these combinations of wavelengths, all the other hues associated with the visible spectrum can be created. They are additive because they are based on the *additive color model*, in which we begin with black, and then add light combinations to create different hues (Figure 10.5). Televisions and computer monitors are also built upon this model, with red, green, and blue phosphors built into each pixel of the screen as a triad. When individual electron beams produced by the hardware's color guns strike the phosphors, red, green, and blue lights are created and combined to create color displays.

Remote sensing and GIS software can take this process a step further by compiling digital images from multiple spectral bands. Some sensors, for example, are designed to sense multiple spectral bands at the same time. In the case of the visible spectrum, this typically means collecting reflected light for the red, green, and blue spectral bands individually. These separate bands can then be displayed individually in grayscale, or combined to create a true color composite. For example, to examine an individual band, we assign that band of

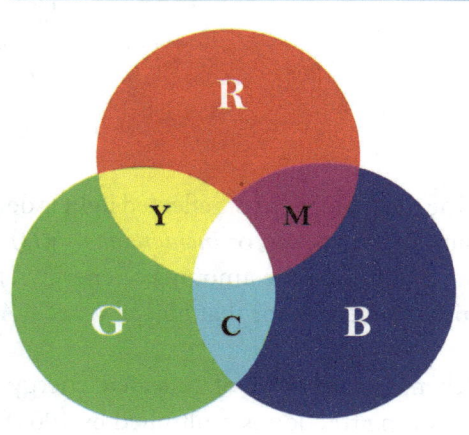

FIGURE 10.5
Additive color model (R=red, G=green, ·B=blue, C=cyan, M=magenta, Y=yellow). Source: Elisabeth Nelson

reflectance values to each of the three color guns. Since each of the color guns creating the image reflects the same amounts of light for each pixel in the image, a grayscale picture of the region is created. This image represents how the area looks when viewed just through the specific lens of that wavelength (Figure 10.6).

In creating the *true color composite*, we assign the red spectral band to the red color gun, the green spectral band to the green color gun, and the blue spectral band to the blue color gun. The software then displays the data from each band using the hues that we would normally associate the features with, and when it combines the different amounts of light reflectance from each hue on the screen, a rendering of the area that looks like a normal color photograph appears. If a black and white, or *panchromatic composite*, is preferred, reflected light for the area would be collected via a panchromatic band, which is a wider band sensitive to the entire range of the visible spectrum. In this case, the one band of panchromatic data would be loaded into each of the color guns; since each gun is reading the same set of reflectance values, the image produced is black and white image. The difference between this image and one of a single band is that the panchromatic covers the entire visible spectrum, mimicking what our eyes would see if we only saw in black and white.

FIGURE 10.6
Displaying an individual spectral band in grayscale. Pixel values are the same for each color gun, resulting in a series of gray shades, such as (80, 80, 80) for the highlighted pixel, representing reflectance values.
Image: U.S. Geological Survey, Department of the Interior.
Source: Elisabeth Nelson

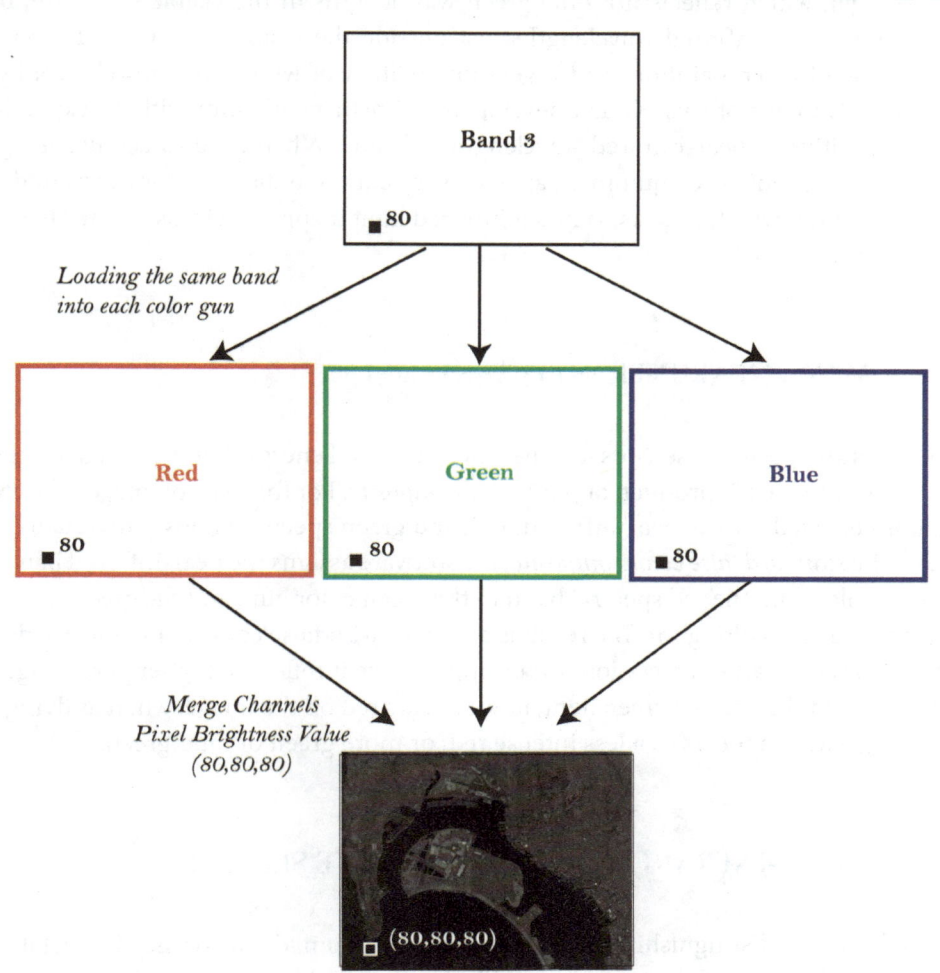

The Near-Infrared Spectrum

Working within the visible spectrum produces images that mimic our vision when color composites are the product, but the electromagnetic spectrum contains many additional wavelengths that we as humans can't process visually. Just on either side of the visible spectrum, for instance, are shorter ultraviolet wavelengths (3 – 0.4 μm) and longer near-infrared wavelengths (0.7 – 1.3 μm). While our eye-brain system does not process either of these, we do have cameras and digital sensors that provide access to them and allow us to 'see' them. The development of sensors to accommodate this extended wavelength range has its roots in World War II and the need, during that time, to be able to detect camouflaged military equipment from the air.

As this equipment was either painted green or covered with cut, green vegetation, its spectral signature in the visible spectrum was very similar to other vegetation, causing it to blend in on panchromatic and true color composite images. There is a difference between live vegetation and camouflaged equipment, though. Live vegetation contains chlorophyll, which reflects not only green wavelengths in the visible spectrum, but also longer near-infrared wavelengths just outside the visible spectrum. The same is not true of green paint or dead vegetation, neither of which contains chlorophyll. To take advantage of this, Kodak developed a panchromatic film with an expanded range sensitive to near-infrared wavelengths of light. When used to acquire images of an area, camouflaged equipment appears very dark and distinct from surrounding live vegetation, which reflects so much infrared light it appears almost white (Jensen, 2000; Scott, 1997).

DISPLAYING NEAR-INFRARED BAND DATA

Scientists also designed sensors to collect near-infrared energy that they used to create a *color infrared (CIR)* product, or *false color composite*. For this type of image, reflected light is collected for the near-infrared, red, and green spectral bands individually. To create the *standard false color composite*, the software assigns the near-infrared band to the red color gun, the red spectral band to the green color gun, and the green spectral band to the blue color gun. The result is an image with false colors, or colors we don't expect to see. Healthy vegetation, for example, since it reflects a higher percentage of infrared light than red or green light, now appears red on the image, whereas dying or dead vegetation appears as a less intense red, or more green or blue-green.

USING NEAR-INFRARED BANDS IN REMOTE SENSING

In addition to distinguishing healthy vegetation from dying or dead vegetation, the infrared region of the spectrum is also useful for identifying different types of vegetation that would appear too similar to tell apart on a traditional color image. The classic example that illustrates this idea uses deciduous and coniferous forests (Figure 10.7). On a traditional panchromatic or true color image, the two types of trees would appear similar in color and tone, as they both reflect approximately the same amount of visible light. Examine them using the infrared portion of the spectrum, however, and deciduous trees will appear much lighter due to the composition of the leaves; conifers, on the other hand, will appear much darker (Avery and Berlin, 1992).

FIGURE 10.7
Comparing panchromatic (top) and black and white infrared (bottom) imagery for distinguishing deciduous trees from conifers. Photos: *U.S. Geological Survey, Department of the Interior.*

Beyond vegetation, scientists use the near-infrared band to highlight land/water contrasts and differences in soil moisture. Land/water contrasts and soil moisture differentiation are common applications because water absorbs most near-infrared wavelengths, creating very dark representations of these features in comparison to dry land, which appears lighter because more infrared wavelengths are reflected (Avery and Berlin, 1992).

Digital Remote Sensing

While traditional photographic systems sense the ultraviolet, visible, and near-infrared spectral bands, atmospheric windows beyond these bands are only accessible through non-photographic technologies. During and following World War II, scientists were busy developing *electronic detection systems* capable of detecting radiant energy for any number of spectral bands. Using electronic detecting elements, *multispectral scanners* were designed to sense more spectral bands, in narrower bandwidths, over a much greater range of the electromagnetic spectrum. Extending the range from 0.3 to ~14 μm, scientists can now access not only the ultraviolet, visible and near-infrared bands, but also the mid-infrared and thermal infrared spectral regions (Lillesand, et al., 2004). When an area is imaged in this way, with multiple separate images for each band, it is called *multispectral imagery.*

Multispectral scanners use electronic detecting elements to record reflected light from the ground; this light is then converted to an electrical charge, which is proportional to the amount of energy the detector element received. Elements that record higher amounts of energy will display on-screen as areas lighter in tone or value than those that record lower amounts of energy. Each band sensed provides an energy reading for the same area, but the reflectance values recorded differ, as they are specific to the individual band (Figure 10.8). The bands of energy collected can be displayed and interpreted individually, or combined into composites by loading them into the red, green and blue color guns to view the merged bands on-screen.

A good example of how this technology expanded our ability to detect reflected energy in other parts of the electromagnetic spectrum is to examine how we harness it to explore *thermal infrared energy.* In this spectral region, the amount of energy

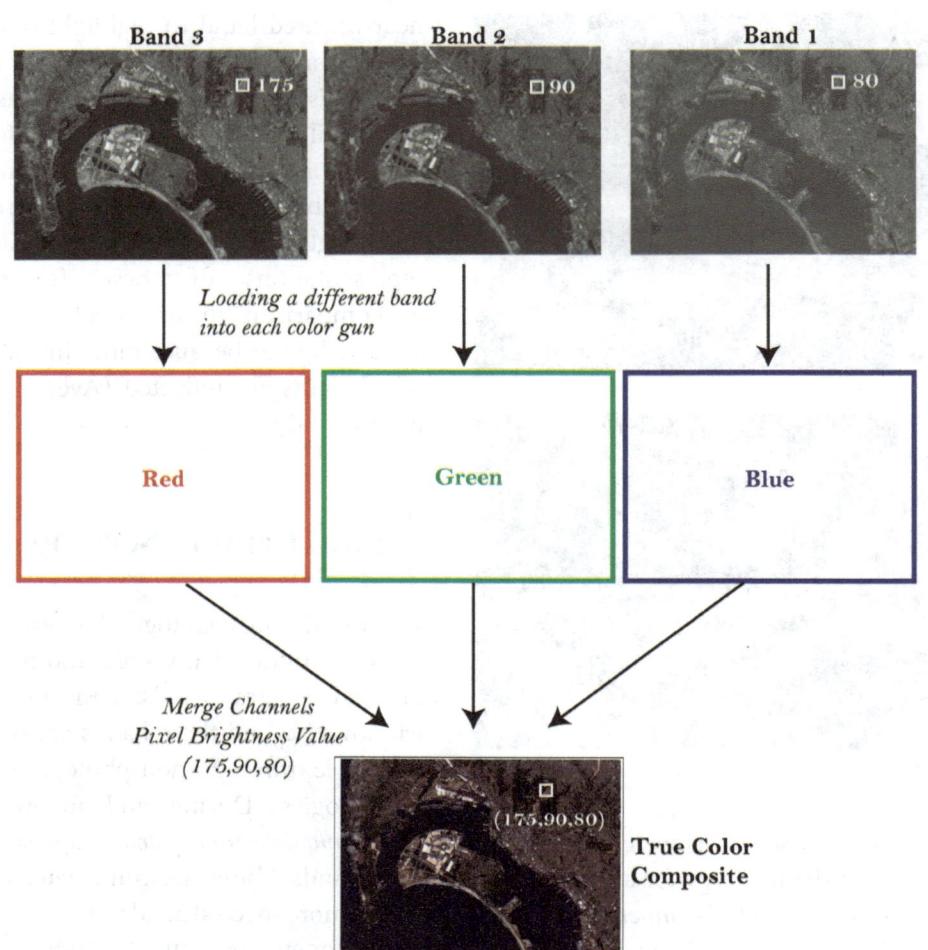

FIGURE 10.8

Displaying three different spectral bands to create a true color composite. Pixel values differ for each band, resulting in different colors, such as (175, 90, 80) for the highlighted pixel, representing reflectance values. Images: *U.S. Geological Survey, Department of the Interior.*
Source: Elisabeth Nelson

emitted by objects correlates with their surface temperature, so warmer objects will appear lighter than cooler objects. Originally conceived for military applications, this spectral band is detectable both day and night. It is also useful for energy loss studies in housing and building designs (Dickinson, 1979).

CHARACTERISTICS OF DIGITAL IMAGERY

With the advent of electronic systems, the products of remote sensing expanded and new terms came into use. Where the principal product of traditional remote sensing is the aerial photograph, electronic systems produce images, which have the capability of being displayed digitally. The term *photograph*, then, refers to any image or picture-like representation detected and recorded via film. *Image* itself is a broader term, meaning any picture-like representation represented digitally, including traditional photographs that have been scanned and saved in digital format (Aronoff, 2005).

Within the electronic scanning systems, images are stored as a grid of equally sized square cells called *pixels*, short for picture elements. Each cell holds a number that represents a brightness value detected for that pixel, where the value is an average of the reflected radiation for all features within that cell. The detail that can be resolved with these systems is dependent, in large part, on the size of the pixel. Any object smaller than the pixel size will not be resolvable. Electronic systems vary in their resolutions, with four different types of resolution playing key roles in determining the final quality of any image for a given application. Understanding these resolutions will help you select data that is best suited for solving your particular problem.

FIGURE 10.9
Comparing spatial resolutions of Landsat imagery: a) 80 m, b) 30 m. Images: *U.S. Geological Survey, Department of the Interior.*

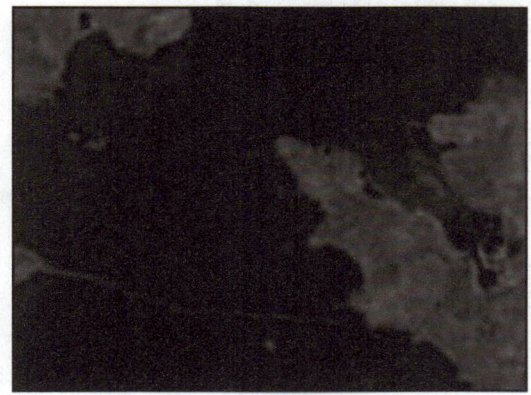

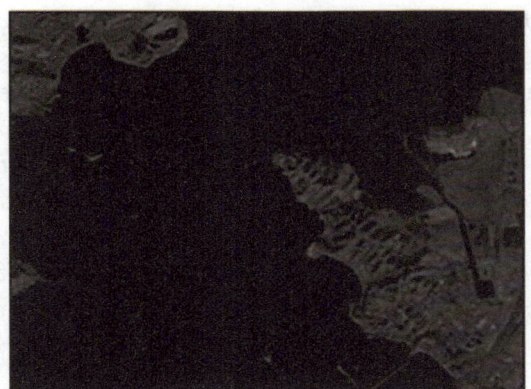

The type of resolution related to an image's pixel size, for example, is its *spatial resolution*. The spatial resolution of a sensor refers to the smallest sized feature you can expect to identify on its corresponding images. The smaller an image's pixels, the more detail you can expect to see and the higher (or finer) the sensor's spatial resolution (Figure 10.9). Larger pixel widths, such as 1 kilometer, are useful for broad land cover studies and for identifying major water bodies, agricultural regions and forested lands. A pixel width of 30 meters, on the other hand, has much more detail and can be used to examine crop patterns and urban areas, as well as to discriminate between vegetation types and identify major roads and airports (Aronoff, 2005).

Sensors and their images also have a *spectral resolution*, which describes the number and width of the spectral bands recorded for each image detected (Table 10.1). These bands may also be referred to as *channels,* and they can be narrow or broad, depending on the sensor's specifications and the applications for which it was designed to detect imagery. For example, a panchromatic band is sensitive to the entire visible spectrum, making it a broader band or channel than a one just sensitive to the red visible band. Because only one storage location is needed to record the reflectance value generated for each pixel, panchromatic bands typically are associated with higher or finer spatial resolutions, which require more storage because the image is divided using a finer grid of smaller pixels.

A sensor composed of three narrower spectral bands, one sensitive to the red visible band, one to the blue visible band, and one to the green visible band, would require three separate storage locations for each pixel in the image. This increase in storage requirements for the spectral resolution is typically offset by decreasing the image's spatial resolution. The result is more flexibility spectrally – you can view any one narrower band or combine any of the image bands to create different color composites, but the resulting image typically will not have the same level of detail as one generated from a broader band.

	LANDSAT 4-5 MSS		LANDSAT 7 TM
Wavelength (mm)			
Bands			
1	0.5 – 0.6		0.45 – 0.52
2	0.6 – 0.7		0.52 – 0.60
3	0.7 – 0.8		0.63 – 0.69
4	0.8 – 1.1		0.76 – 0.90
5			1.55 – 1.75
6			10.40 – 12.50
7			2.08 – 2.35

TABLE 10.1
Spectral resolutions for Landsat MSS and TM sensors.

The number of brightness levels recorded for each pixel is a measure of *radiometric resolution* (Figure 10.10). This type of resolution is also tied to the storage capabilities of the sensor. Electronic devices, at their simplest, work as on/off switches, where each pixel would either detect energy reflectance (and code it as a 1) or detect no energy reflectance (and code it as a 0). These values are called *bits* in the computer world; their foundation is base 2 mathematics (as opposed to base 10, with which we typically work) and they form the basis for storing information in a computer. The result in this case is a black and white image; differences in the percentages of reflectance don't matter because there isn't any way to record that detail with only an on/off mentality.

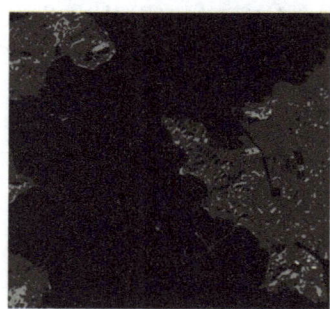

FIGURE 10.10
Comparing radiometric resolutions: 2-bit (top), 4-bit (middle), 8-bit (bottom). Images: *U.S. Geological Survey, Department of the Interior.*

If, however, we design the system so that each pixel receives two bits of memory, then each pixel could store 1 of 4 different values. This increases the radiometric resolution and allows the sensor to record finer variations in reflectance values. A typical format today would be to allocate either 8 or 16 bits to each pixel to capture a range of brightness values useful for analytical processes. Eight bits, for example, will allow each pixel to record up to 256 different brightness levels, which mimics the range of gray tones our eyes can perceive. Like spectral and spatial resolution, radiometric resolution also affects the file size of an image. Take an image covering 60 km x 60 km. If the spatial resolution is 20 m, then the image must be divided into 3000 pixels per side (3000 x 20 = 60,000 m or 60 km). That's 9 million pixels to cover the area (3000 x 3000) completely. If we record each band using 8 bits (1 byte) of computer memory, we will need roughly 9 Mb of storage to store this file on our computer.

The last type of resolution to consider is *temporal* resolution. Temporal resolution describes how frequently an image is acquired of an area. Sensors with higher temporal resolutions revisit areas more frequently. High temporal resolutions are useful for crop monitoring studies and studies examining change detection of an area.

Satellite Systems

Spaceborne platforms for remote sensing got their start in the United States with the launching of *American Explorer I*, our response to the Soviet's launching of *Sputnik-I* in 1957. By 1960, we had generated our first space photographs, produced by using cameras and film aboard a spy satellite in the Corona program. Scientists designed the system to return the film via capsule and parachute, and then destroyed the satellite following the mission. These systems quickly moved to all digital components and stable, recurring platforms that provided better image resolutions, longer missions, and even the ability to reposition the satellites (Aronoff, 2005).

Satellite imagery offers several advantages over standard aerial photography, but it does not replace it. For example, satellite or spaceborne platforms provide continuous, global data acquisition (increasing temporal resolution) and access to a more standardized product, both of which make temporal comparisons used in change detection studies easier to conduct. The scale of the imagery, however, is small, and provides minimal capability for the stereoscopic viewing required in large-scale topographic applications (Thrower, 1996). It also provides the same projective view as standard aerial photography, resulting in similar distortion properties, although these become slightly more complicated as they can vary depending on the type of orbit the satellite follows.

Some satellites, for example, are placed in *geostationary orbits*. These orbits are designed so that the satellite rotates at the same speed and in the same direction as Earth, keeping it over the same point on the equator at all times; this allows it to monitor the same area continuously (Figure 10.11). Geostationary orbits are most typical of weather satellites, whose primary purpose is to monitor patterns of cloud cover at high altitudes. The resulting imagery is global in scale, with the same geometric distortions common in small-scale maps. Other satellites are placed in *Sun-synchronous, near-polar orbits*, in which the satellite rotates westward as the Earth rotates eastward, allowing it to cross different locations at approximately the same time each

FIGURE 10.11
Geostationary orbit.
Courtesy of NASA/JPL-Caltech.

day (Figure 10.12). The result is less frequent imaging of the area and additional distortion due to rotation, but the images that are produced all have consistent sun angles for easier interpretation over time and more detailed views because these orbits are typically lower in altitude than the geostationary ones. Satellite systems, then, serve as complementary approaches to traditional aerial photography, expanding the overall scope of remote sensing (Monmonier and Schnell, 1984).

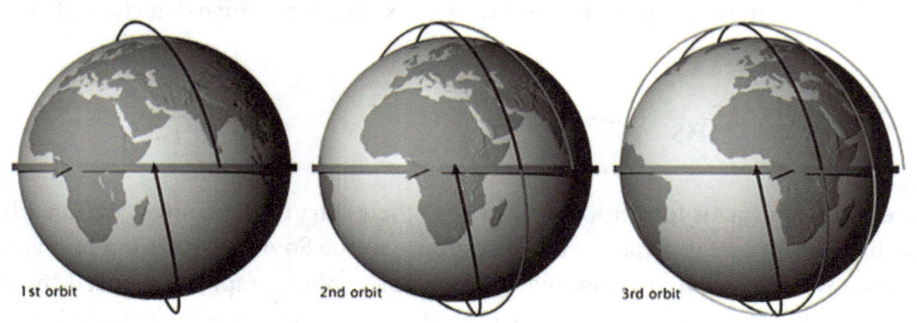

1st orbit   2nd orbit   3rd orbit

FIGURE 10.12
Sun-synchronous, near-polar orbit. *Courtesy of NASA/JPL-Caltech.*

Different satellite systems are designed to address different remote sensing applications; sensor resolutions, coverage, frequency, cost, and speed of image delivery are all variables that are addressed within this context. Low-resolution systems, for example, monitor global conditions and are useful for assessing crop conditions and classifying regional land covers and land uses. The first of these satellites to be dedicated to a civilian application, meteorology, was *TIROS-1* (Television and Infrared Observation Satellite) which was launched in 1960. Its resolution was coarse, but it highlighted the potential of satellites to influence weather mapping. The technology behind this satellite series also formed the foundation for the United States' first earth resources satellite series, *ERTS* (Earth Resources Technology Satellites). This series, launched in 1972, marked the first continuous surveillance of Earth using satellites. Renamed *Landsat* in 1975, it has the backing of both the United States Geological Survey (USGS) and the National Aeronautical and Space Administration (NASA). Their support has allowed earth resources remote sensing to develop at a rapid rate (Thrower, 1996).

Landsat, a medium resolution series, offered the first systematic repetitive coverage of Earth with enough detail to be useful for a variety of applications. The earlier satellites in this series used a multispectral scanner sensor (MSS) that operated in 4 spectral regions similar to color infrared film photography: the visible red, green, and blue bands, plus a near-infrared band. With a Sun-synchronous, near-polar orbit at 900 km and a maximum spatial resolution of 80 m on earlier satellites, the final product from these systems were higher quality images with significantly reduced distortion. Images could be acquired every 18 days for the same area. Interests from areas such as agriculture, forestry, geology, archaeology, and urban and regional land use helped guide the development of later sensor designs, including choice of bandwidth and placement within the electromagnetic spectrum.

Today, the Landsat program is managed by a combination of NASA, NOAA, and the USGS, who operate the program under the *open skies policy*, allowing them to provide the data free to the public worldwide. Images from Landsats 1-3 (launched in 1972, 1975, and 1978) were digital products, and of high enough quality that they prompted the development of *digital image processing*, the development of computer procedures and algorithms to enhance visual image interpretation. The launching of Landsat 4 in 1982 failed, but Landsat 5, launched in 1984, added a new sensor to the mix, the *Thematic Mapper™*. This sensor offered a 30 m spatial resolution and increased the spectral resolution from 4 bands (MSS) to 7, covering the visible, near infrared and thermal infrared spectral regions, with the thermal band having a spatial resolution of 120 m. Repeat coverage improved from 18 days to 16. Like Landsat 4, Landsat 6 (1993) failed. Landsat 7, however, launched in 1999, again offering an improved sensor, the *Enhanced Thematic Mapper (ETM+)*. This sensor offered 7 spectral bands at a 30 m spatial resolution, along with a 15 m panchromatic band, a 60 m thermal infrared band and improved radiometric resolution. This system malfunctioned in 2003, leaving only Landsat 5 still functioning (NASA, 2013). Landsat 8 launched in February 2013 adding a second working platform in the Landsat series of satellites. The Landsat 8 satellite provides enhancements from previous Landsat instruments. Two additional spectral bands were added; a deep blue visible channel (band 1) and a shortwave infrared channel (band 9). Band 1 is designed to help with coastal zone and water resources investigation while band 9 is designed to detect cirrus clouds (Table 10.2).

TABLE 10.2
LANDSAT 8 Operational Land Imager (OLI) and Thermal Infrared Sensor (TIRS)

Bands	Wavelength (micrometers)	Resolution (meters)
Band 1 – Coastal aerosol	0.43 – 0.45	30
Band 2 – Blue	0.45 – 0.51	30
Band 3 – Green	0.53 – 0.59	30
Band 4 – Red	0.64 – 0.67	30
Band 5 – Near Infrared (NIR)	0.85 – 0.88	30
Band 6 – SWIR 1	1.57 – 1.65	30
Band 7 – SWIR 2	2.11 – 2.29	30
Band 8 – Panchromatic	0.50 – 0.68	15
Band 9 – Cirrus	1.36 – 1.38	30
Band 10 – Thermal Infrared 1	10.60 – 11.19	100
Band 11 – Thermal Infrared 2	11.50 – 12.51	100

From 1972 to 1986, the United States was the only country with a satellite-based remote sensing system that was virtually global in scope. In 1986, France joined us with the launching of their system, *SPOT* (Système Pour l'Observation de la Terre), which offered a 10 m spatial resolution in its panchromatic band and a 20 m spatial resolution in the visible and near infrared bands. Also unique to their satellite at the time was advanced optic capability, including not only nadir viewing but also off-nadir, or side viewing, which enabled France to produce stereopairs from their imagery more easily. SPOT has also seen an evolution in technology including adding a 20 m mid-infrared band on SPOT 4 in 1998 and a 5 m panchromatic band on SPOT 5 in 2002 (Astrium, 2013).

Even with France's contributions in the 1980s, and other countries and entities like India and the European Union launching systems in the 1990s, Landsat was still the most detailed publicly available imagery until 1999. In that year, EOSAT (now Space Imaging) launched the first commercial high-resolution satellite system, IKONOS. Offering a spatial resolution of 1 m for its panchromatic band, IKONOS imagery gave civilian applications what before had only been available to the military: supreme spatial resolution with an increased capability to revisit areas (11 days) and provide redundant image acquisition. IKONOS was quickly followed by other ventures, like Quickbird by DigitalGlobe, Inc., who tailor their systems to the needs of their clients, with prices varying according to the spatial, spectral and radiometric resolutions requested, as well as the level of geometric correction required. Most offer a panchromatic band, as well as up to four multispectral bands, and unlike government and military systems, most produce oblique imagery using sensors that point to the side. Such capabilities make them quite valuable for monitoring urban infrastructure and providing emergency response to natural disasters as well as evidence for breaking news stories (Aronoff, 2005).

Digital Image Processing

Using these digital images, particularly with other geographic data, requires that we be able to manipulate them digitally. While most of these procedures fall outside the scope of this text, it is important to introduce you to at least a few. Here, we explore image restoration and rectification processes, image enhancement procedures, and the basic concept of image classification.

IMAGE RESTORATION AND RECTIFICATION

Many of the algorithms you will find under this heading are for processes that can also be performed by the sellers of the imagery prior to purchasing it. Correcting degraded data, removing systematic geometric distortions, and changing image geometry are examples of the types of tasks that must be run on raw imagery before it can be used in a GIS. The amount of reflected energy that sensors measure, for

FIGURE 10.13

Georectification: nearest neighbor transformation. Source: Elisabeth Nelson

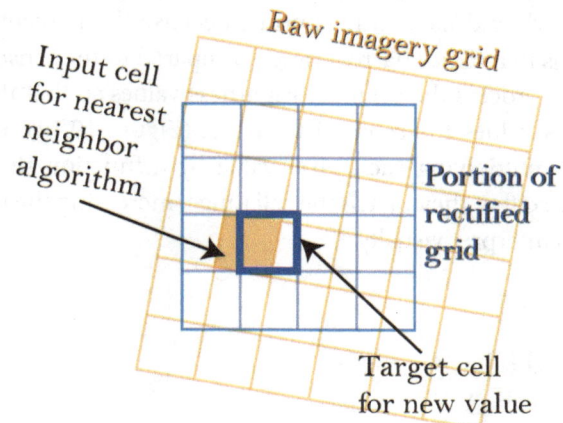

example, is influenced by shadows, haze and other atmospheric conditions. If one of these issues is a systematic influence occurring across an image, algorithms that process the raw imagery can compensate for this.

Raw imagery must also be tied to ground locations to be used in a GIS. *Georectification* is the process of assigning a geographic coordinate to every pixel in an image. Basically a two-step process, georectification begins with a mathematical transformation to calculate the geographic coordinates that will tie the imagery to a map projection used by other GIS data. The most straightforward way of doing this is to identify a set of coordinates in the field or on a reference source (called *ground control points*) and then locate those same points on the image. The transformation algorithm, using the data associated with the original pixels, then creates a new grid of pixels using the desired map projection and calculates new cell values. The simplest algorithm for this part of the process, and the only one that doesn't change the original pixel values, is the *nearest neighbor* algorithm (Aronoff, 2005). This algorithm calculates the pixel centers for the new grid, then finds the corresponding location on the original image and assigns the value of the pixel nearest to that location to the new grid cell (Figure 10.13).

FIGURE 10.14

Contrast stretch: Original image (top), modified image (bottom). Used with permission. Copyright 2018. Purdue University, Multispec.

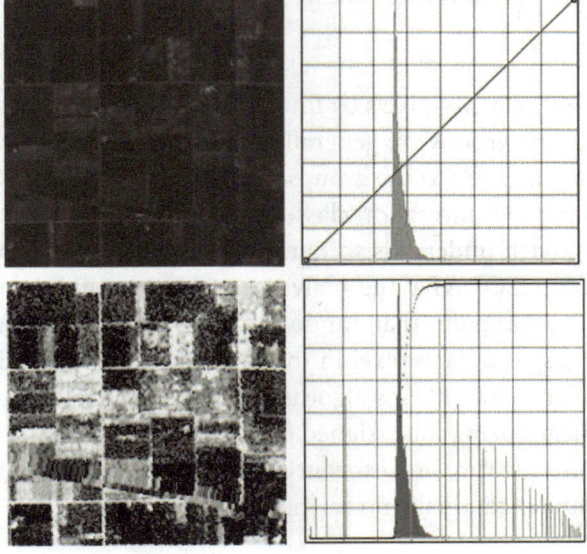

IMAGE ENHANCEMENT

Image enhancement algorithms are used to improve our ability to interpret an image visually. They include processes to manipulate contrast, brightness values, and image sharpness. The point of these algorithms is not to add information, but merely to exaggerate it. For instance, many satellite images originally look quite dark and have little contrast because the percentages of energy reflected to the sensor is in a rather narrow range compared to the sensor's capabilities. It makes sense, then, to process the original brightness values so that they take up the full range of brightness values the sensor can detect (Figure 10.14). *Contrast stretch* algorithms do this by multiplying each pixel value by a function that stretches the original range of values so that they match the full range, increasing the image's contrast and making it easier to interpret visually.

IMAGE CLASSIFICATION

Image classification is popular because it allows scientists to classify pixels by brightness value into useful groups. There are two basic approaches for this: *supervised* and *unsupervised* classification. The key is that information classes are not known prior to these processes; they have to be generated from the original pixel values. In the supervised approach, the analyst defines the classes the algorithm will extract by selecting training areas on the original imagery. These are areas on the image the analyst can identify by experience, fieldwork, or use of other higher resolution imagery. The computer then generates a spectral signature for these areas, plots the correspondences between bands on a scatterplot, and then uses that information to classify all of the pixels in the image.

The simplest means of doing this is by using an algorithm known as the *Minimum Distance to Means*. It compares the spectral value of each pixel to the arithmetic mean for each training class in each spectral band, and then assigns the pixel to the class that minimizes the difference between average pixel value for the class in each band and the pixel's value in each band. The algorithm is simple and fast, but has a tendency to misclassify pixels because different training areas have different ranges of spectral reflectances, which can affect the arithmetic mean calculated for the class (Aronoff, 2005). The final accuracy of the classification is then assessed using independent data for verification purposes.

Unsupervised classification algorithms classify pixels by finding statistical groupings in the image's data values. The resulting groups are generally well defined spectrally, but the analyst must assign meaning or labels to the groups, which can be difficult if the group doesn't correspond well to the information classes with which the analyst is working. The most common approach under this scenario is to use an algorithm called *ISODATA* (Aronoff, 2005). With ISODATA, the analyst specifies the number of clusters for the computer to find. The computer than randomly seeds a cluster mean for each cluster in the multispectral space. Each pixel in the image is then assigned to one class using the Minimum Distance to Means algorithm. When complete, the computer recalculates the cluster means using the assigned data, and then reclassifies the image using these new means. This iterative process continues for either a specified number of rounds or until the computer can make no further adjustments.

References

Aronoff, Stan. 2005. *Remote Sensing for GIS Managers.* Esri Press: Redlands, California.

Astrium. 2013. *Technical Information about the SPOT Satellites. http://www.astrium-geo.com/na/1240-spot-technical-information* Last accessed 11 Jan 2013.

Avery, T.E. and G.L. Berlin, 1992. *Fundamentals of Remote Sensing and Air Photo Interpretation.* 5th ed. Macmillan Publishing Company: New York, New York.

Campbell, J.C. 1984. *Introductory Cartography.* Prentice-Hall, Inc.: Englewood Cliffs, New Jersey.

Campbell, J.B. 1996. *Introduction to Remote Sensing.* Guilford Press: New York, New York.

Dickinson, G.C. 1979. *Maps and Air Photographs.* 2nd ed. Edward Arnold, Ltd.: London, England.

Jensen, J. R. 2000. *Remote Sensing of the Environment: An Earth Resource Perspective.* Prentice-Hall, Inc.: Upper Saddle River, New Jersey.

Lillesand, T. and R.W. Keifer, 2007. *Remote Sensing and Image Interpretation.* 6th ed. John Wiley & Sons: New York, New York.

Monmonier, M. and G.A. Schnell, 1984. *Map Appreciation.* Prentice-Hall, Inc.: Englewood Cliffs, New Jersey.

NASA. 2012. *Imagine the Universe!* Retrieved from http://imagine.gsfc.nasa.gov/docs/dictionary.html Last accessed 01 Jan 2013.

NASA. 2013. *LANDSAT.* Retrieved from http://landsat.gsfc.nasa.gov/ Last accessed 05 Jan 2013.

NASA. 2012. *Remote Sensing.* Retrieved from http://earthobservatory.nasa.gov/Features/RemoteSensing/remote.php Last accessed 05 Jan 2013.

Natural Resources of Canada. 2008. *Electromagnetic Radiation.* http://www.nrcan.gc.ca/earth-sciences/geography-boundary/remote-sensing/fundamentals/1784 Last accessed 02 Jan 2013.

Nave, C.R. 2012. *Hyperphysics.* http://hyperphysics.phy-astr.gsu.edu/hbase/atmos/blusky.html Last accessed 05 Jan 2013.

Rabenshorst, T.D. and P.D. McDermott, 1989. *Applied Cartography: Introduction to Remote Sensing.* Merrill Publishing Company: Columbus, Ohio.

Shellito, B. 2012. *Introduction to Geospatial Technologies.* W.H. Freeman and Co.: New York.

Thrower, N.J.N. 1996. *Maps and Civilization: Cartography in Culture and Society.* University of Chicago Press: Chicago, Illinois.

Exercise 10

Exploring Satellite Imagery

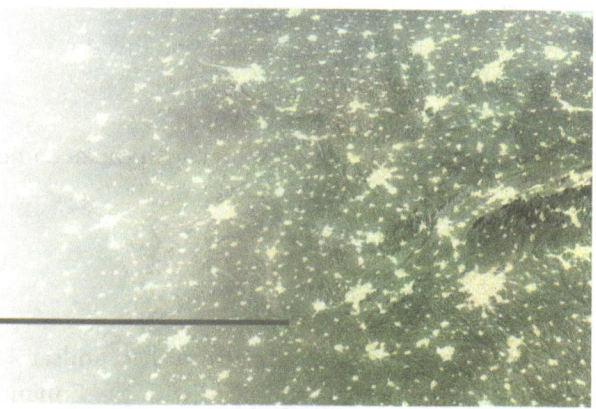

In Exercise 10, you will learn to load and manipulate spectral bands associated with multispectral satellite imagery.

OBJECTIVES

- Load satellite bands into remote sensing software to display imagery
- Manipulate imagery scale and identify locations by coordinates
- Create and interpret different color composites
- Create spectral profiles and compare brightness values of different features

SOFTWARE INFORMATION

This lab uses **MultiSpec,** a freeware multispectral image data and analysis system. If the software is not available in your geospatial lab, it may be downloaded for free at https://engineering.purdue.edu/~biehl/ MultiSpec/

DATA

The data for the exercises is available from the Kendall Hunt Student Ancillary site. See the inside front cover for access information. You may also be directed to download the data from a different location by your instructor.

1. Download the data as instructed by your instructor
2. Save the file to a location where you have read/write privileges (USB key, home computer, class server)

The file you just saved is in a compressed (*.zip) format, and was created using **7-zip** freeware (http://www.7- zip.org). To use the data for this exercise, you must decompress it using the same software.

3. Locate the zipped file that you saved
4. Right-click the file
5. Choose *7-Zip - Extract Files*
6. Click *OK* to create a folder with the decompressed data

Displaying Imagery

The assignment uses imagery from the Landsat 7 satellite. Launched as a joint project between NASA and the USGS, Landsat 7 is equipped with the Enhanced Thematic Mapper Plus (ETM+) multispectral sensor. ETM+ collects eight bands of data ranging from the visible to the thermal infrared. More information is available at the following website: http://landsat.gsfc.nasa.gov/about/etm+.html

Each of these bands may be visualized by loading them individually or in any combination into the color guns of your monitor using remote sensing/GIS software. Different combinations lend themselves to different interpretations of the landscape, as we will see below.

QUESTION 10.1

Using your textbook and/or lecture notes match each of the bands collected by ETM+ to the section of the electromagnetic spectrum sensed.

Band 1 (0.45 - 0.515 mm)	_____	A. Mid infrared
Band 2 (0.525 – 0.605 mm)	_____	B. Panchromatic/Visible
Band 3 (0.63 – 0.69 mm)	_____	C. Blue
Band 4 (0.75 – 0.90 mm)	_____	D. Mid Infrared
Band 5 (1.55 – 1.75 mm)	_____	E. Green
Band 6 (10.4 – 12.5 mm)	_____	F. Thermal Infrared
Band 7 (2.09 – 2.35 mm)	_____	G. Red
Band 8 (0.52 – 0.9 mm)	_____	H. Near Infrared

1. Start MultiSpec
2. From the *File* menu inside the application select *Open Image*
3. Navigate to your *Chapter10* folder
4. Select *GSO_ETMplus_7bands.tif* and click *Open*
5. A dialog box (Figure 10.1) requesting information on display parameters will appear. Notice that the default bands or channels to be displayed are bands 4, 3, and 2.

FIGURE 10.1
Used with permission.
Copyright 2018. Purdue
University, Multispec.

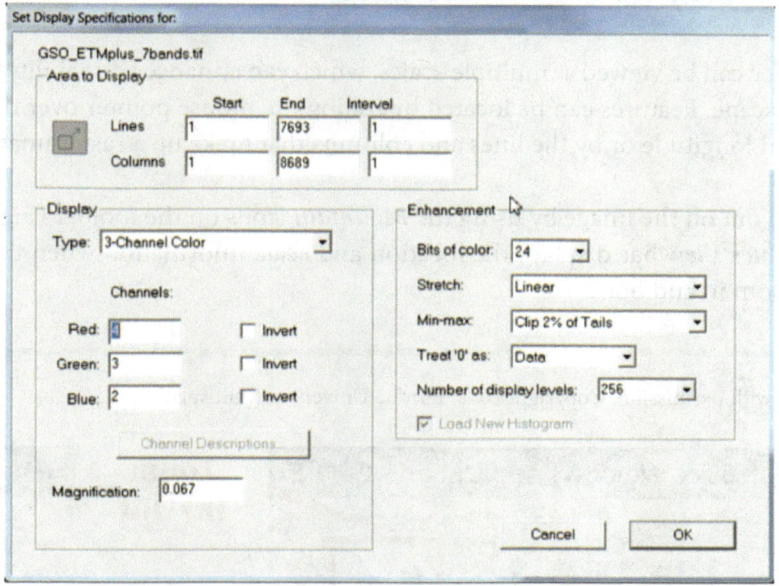

6. Click *OK*
7. If prompted, click *OK* again

QUESTION 10.2

When bands 4, 3, and 2 of the ETM+ sensor are respectively assigned to red, green, and blue, what type of image results?

a. True color composite
b. False color composite
c. Panchromatic
d. Thermal infrared

QUESTION 10.3

Areas that appear bright red on this image are reflecting energy primarily from which part of the electromagnetic spectrum?

a. Infrared
b. Blue visible
c. Red visible
d. Green visible

Identifying Imagery Locations and Scale

Imagery in MultiSpec can be viewed at multiple scales, which can enhance your ability to detect and identify features in the landscape. Features can be located by rolling the mouse pointer over the image and locating them by latitude and longitude or by the lines and columns that make up a raster image.

1. Zoom in and out on the image by using the *Mountain* icons on the toolbar (Figure 10.2).
2. The *Coordinates View* bar displays the location and scale information when the image is active and when you zoom in and out.

FIGURE 10.2 Used with permission. Copyright 2018. Purdue University, Multispec.

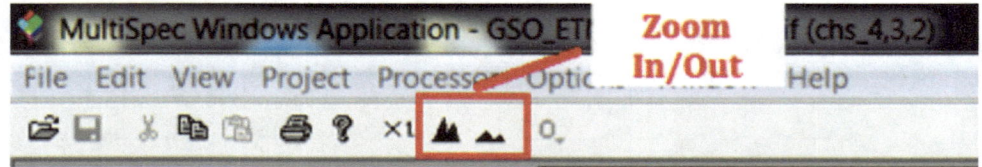

3. The scale of the image will show in the bar (Figure 10.3); keep zooming in until the scale reads between 1:10,000 and 1:16,000. You should be able to see individual pixels at this scale.

FIGURE 10.3

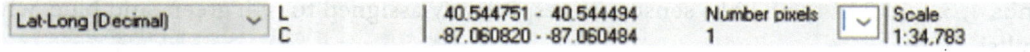

The default display unit is Latitude/Longitude in Decimal Degrees. As you scroll your cursor over the image, numbers in the coordinates view window change.

4. Find an interesting looking pixel. Move your cursor over it and double-click. The *Coordinates View* bar has captured numbers related to the pixel you have selected. Note: Once you have double-clicked the pixel, MultiSpec should keep the image centered on it as you zoom in and out.

QUESTION 10.4

You have selected a pixel. Record the number ranges that were captured in the *Coordinates View* bar.

_____ - _____

_____ - _____

These numbers represent the size of the pixel in:

a. State Plane coordinates
b. UTM coordinates
c. Latitude/Longitude
d. Mercator coordinates

5. MultiSpec also gives you the ability to navigate to a specific pixel. In the drop-down window on the left side of the *Coordinates View* bar (Figure 10.4), click the *Down Arrow* button and change the coordinates from Lat-Long (Decimal) to Lines-Columns.

FIGURE 10.4
Multispec Coordinates View Bar showing units of measurement.
Used with permission.
Copyright 2018. Purdue University, Multispec.

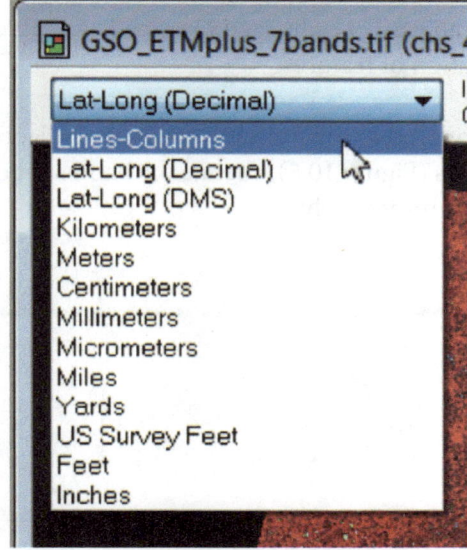

Individual pixels can be found by navigating through the grid. Grid locations are laid out and numbered from left to right (Columns) and from top to bottom (Lines).

Study the numbers to the right of the *Coordinates View* bar as you move your cursor vertically and horizontally through the image. Notice that the numbers decrease as you move up and left and increase as you move down and right.

Manipulating and Interpreting Color Composites

Imagery created from bands 4, 3, and 2 of ETM+ tend to have several distinct characteristics:

- Water does not reflect much energy from any of these bands
- Urban features reflect a lot of energy from bands 3 (red wavelengths displayed using the monitor's green color gun) and 2 (green wave lengths displayed using the monitor's blue color gun)
- Healthy vegetation reflects a tremendous amount of energy from band 4

QUESTION 10.5

Scroll over the image and double-click once in the general area of each pair of coordinates specified below. Zoom in or out on the image until features become identifiable (approximately 1:80,000). Using the information above, match the objects in the general area of these coordinates with their correct identification (you may need to zoom in and out as you complete this task).

1. Line 3673, Column 2345 _____ Water
2. Line 3692, Column 3740 _____ Airport
3. Line 1744, Column 7339 _____ Urban/City
4. Line 3680, Column 3914 _____ Golf Course

The bands comprising the image can easily be changed to provide a different visualization of the landscape, which can help enhance interpretation of features.

1. To change band combinations, go to the *Processor* menu and select *Display Image*.
2. In the dialog box that appears (Figure 10.5), change the band assignments of the red, green, and blue channels to bands 3, 2, and 1 respectively.
3. Click *OK*

FIGURE 10.5
Used with permission. Copyright © 2018 Esri, ArcGIS, ArcMap, ArcCatalog, United States Department of the Census; naturalearthdata.com All rights reserved.

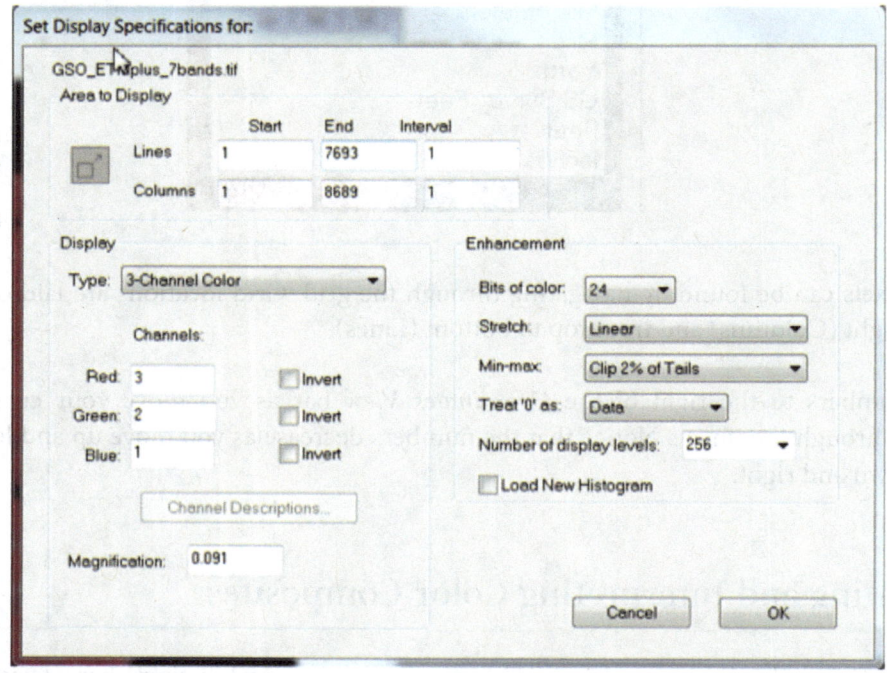

QUESTION 10.6

What type of image results when Band 3 (red wavelengths) is assigned to the red color gun, Band 2 (green wavelengths), and Band 1 (blue wavelengths) to the blue color gun?

a. True color composite
b. False color composite
c. Panchromatic
d. Thermal infrared

Creating and Examining Spectral Profiles

Each pixel in a band holds a brightness value (BV) between 0 and 255, which represents the amount of light in that band reflected from the earth's surface at that point. The values for a pixel can be collected from multiple multispectral bands and graphed, placing bandwidth or channel number along the x-axis and BV along the y-axis. This is called a spectral profile.

1. Go back to the *Processor* menu – *Display Image* and change the bands back to 4,3,2 for red, green and blue
2. Zoom out to see the entire image, then double-click once in the area marked A (Figure 10.6) and zoom in to that region. As you zoom, center your viewer on the area that is bluish-white. You should notice a linear feature (Piedmont Triad International Airport) at a scale of about 1:80,000 or larger, and Greensboro should be to the east (right).
3. Once you have zoomed in far enough to see individual pixels, highlight a bluish-gray pixel on the runway by double-clicking it
4. From the menu bar, click on *Window,* then *New Selection Graph*
5. Resize the chart as needed to view the graph and the information along each axis

FIGURE 10.6
Source: U.S. Geological
Survey, Department of
the Interior

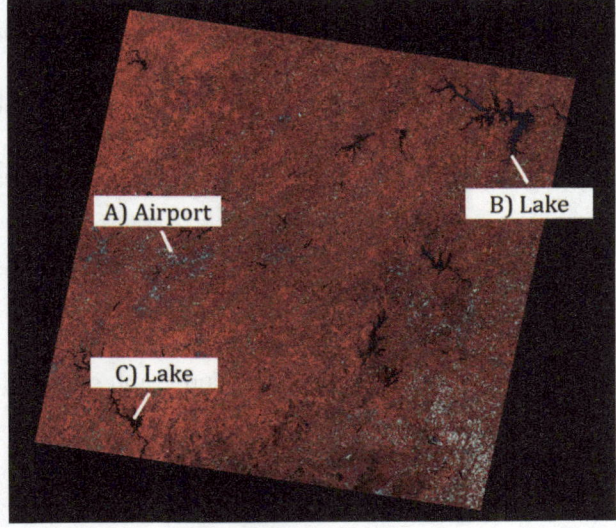

QUESTION 10.7

Using the information found at the top of the chart, record the pixel location for the pixel location from the sample of the airport runway.

Line _____   Column _____

QUESTION 10.8

Record the approximate values from the election graph on the following chart (Figure 10.7). To see the actual values for each band, go to the *Processor* menu, and choose *List Data*. Click *OK*, then click on the text window behind the image to see the values.

FIGURE 10.7

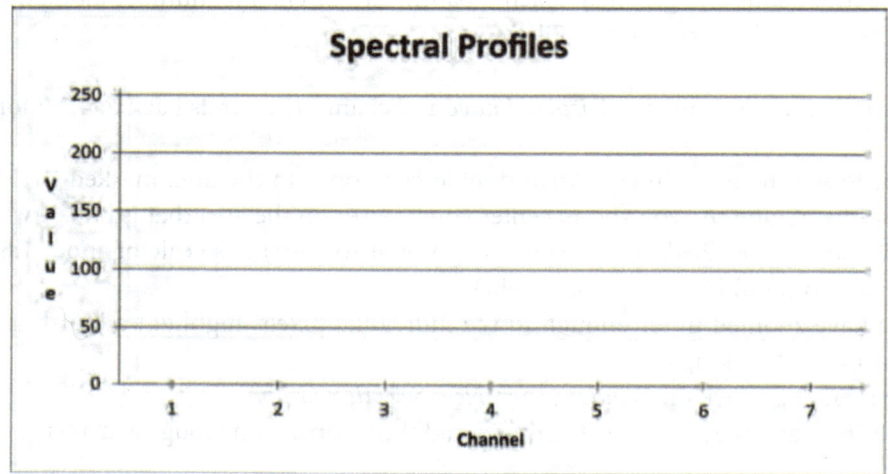

6. Close the selection graph window, and then zoom out (try approximately 1:1600,000) from the runway in Area A (Figure 10.6). To the southwest of the runway, center your display on the lake (Area B: a very dark blue/black object) and double-click.
7. Double-click on one of the lake's pixels to select it, then zoom in on the object.
8. From the menu bar, click on *Window*, then *New Selection Graph*.
9. Resize the chart as needed to view the graph and the information along each axis.

QUESTION 10.9

Record the pixel location and chart the values from the selection graph on the chart (Figure 10.8).

Line _____   Column _____

FIGURE 10.8

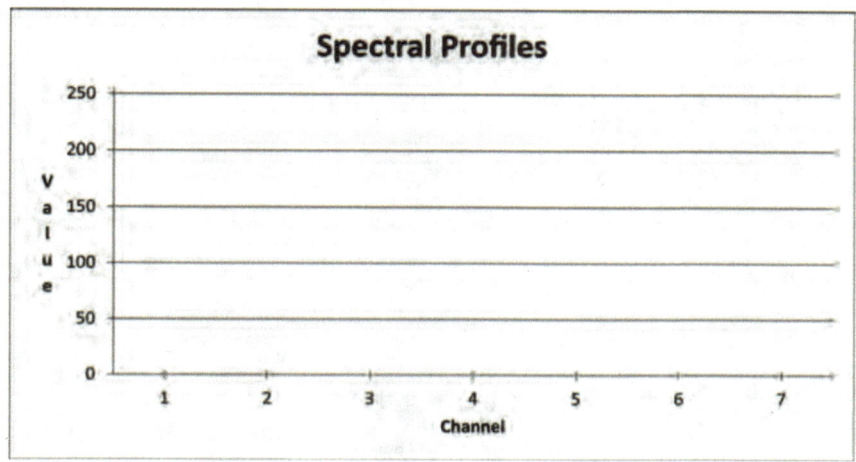

QUESTION 10.10

The lake reflects less Band/Channel 4 energy than the airport.

   a.   True
   b.   False

10. Close the selection graph window, and then zoom out again until you see the are marked C in Figure 10.6
11. Double-click once in this area, and zoom in to the area to the right of the southernmost lake until you reach a scale between 1:15,000 and 1:25,000
12. Navigate to and then double-click on Line 6376, Column 1851
13. From the *Processor* menu, choose *List Data*
14. Click *OK*
15. Click on the text window behind the image to bring it to the front
16. Record the approximate values for band 1 through band 7 for **Pixel 1** in the table

	Pixel 1 (Line 6376, Column 1851)	Pixel 2 (Line 6405, Column 1831)
Band 1		
Band 2		,
Band 3		
Band 4		
Band 5		
Band 6		
Band 7		

17. Plot the spectral profile in the graph shown in Figure 10.9

FIGURE 10.9

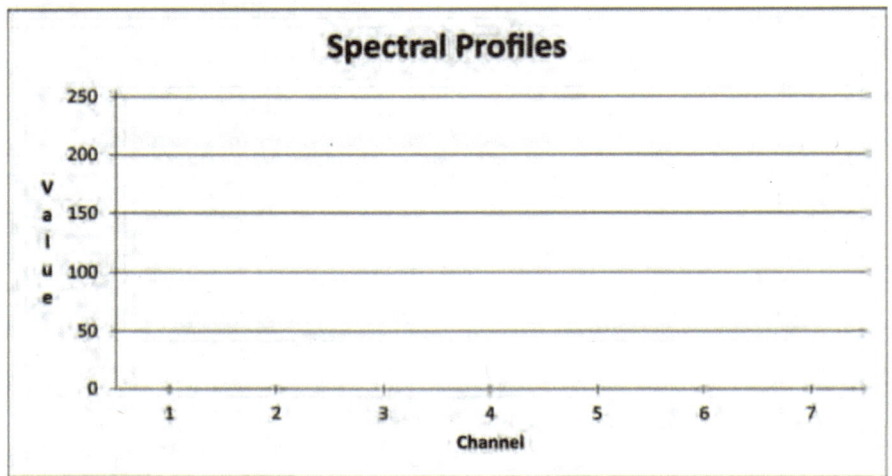

18. Navigate to the second pixel (Pixel 2): Line 6405, Column 1831
19. From the *Processor* menu, choose *List Data*
20. Click *OK*
21. Click on the text window behind the image to bring it to the front
22. Record the values for band 1 through band 7 for **Pixel 2** in previous table where you recorded the values for **Pixel 1**
23. Plot the spectral profile of **Pixel 2** on the chart shown in Figure 10.9 as well (it should overlay **Pixel 1** but use dash lines for **Pixel 2** so that you can distinguish between the two profiles)

QUESTION 10.11

Pixel 1 reflects significantly more Band 4 energy when compared to Pixel 2.

   a. True
   b. False

QUESTION 10.12

Pixels 1 and 2 represent different types of tree species: one represents conifers, and the other represents deciduous.

Based on the spectral signatures that you plotted and the knowledge you have gained in class, which type of tree should be located at Pixel 1?

   a. Deciduous
   b. Conifer

24. Exit all applications